ENVIRONMENTAL REMEDIATION TECHNOLOGIES, REGULATIONS AND SAFETY

GEOPHYSICS

PRINCIPLES, APPLICATIONS AND EMERGING TECHNOLOGIES

ENVIRONMENTAL REMEDIATION TECHNOLOGIES, REGULATIONS AND SAFETY

Additional books in this series can be found on Nova's website under the Series tab.

Additional e-books in this series can be found on Nova's website under the eBooks tab.

ENVIRONMENTAL REMEDIATION TECHNOLOGIES, REGULATIONS AND SAFETY

GEOPHYSICS

PRINCIPLES, APPLICATIONS AND EMERGING TECHNOLOGIES

GEMMA AIELLO
EDITOR

nova publishers
New York

Additional color graphics may be available in the e-book version of this book.

Library of Congress Cataloging-in-Publication Data

Names: Aiello, Gemma, editor.
Title: Geophysics: principles, applications and emerging technologies / editor: Gemma Aiello.
Other titles: Geophysics (Nova Science Publisher's Inc.: 2016)
Description: Hauppauge, New York: Nova Science Publisher's, Inc., 2016. |
Series: Environmental remediation technologies, regulations and safety |
Includes index.
Identifiers: LCCN 2016007430 (print) | LCCN 2016015906 (ebook) | ISBN 9781634848312 (hardcover) | ISBN 9781634848527 (ebook) | ISBN 9781634848527 ()
Subjects: LCSH: Geophysics--Methodology. | Geophysics--Technological innovations. | Geophysics--Italy.
Classification: LCC QC808.5 .G47 2016 (print) | LCC QC808.5 (ebook) | DDC 550--dc23
LC record available at https://lccn.loc.gov/2016007430

Published by Nova Science Publishers, Inc. † New York

CONTENTS

Preface		**vii**
Chapter 1	Environmental Geophysics: Techniques, Advantages and Limitations *Pantelis Soupios and Eleni Kokinou*	**1**
Chapter 2	Application of Innovative Geophysical Techniques in Coastal Areas *Vincenzo Di Fiore, Michele Punzo, Daniela Tarallo and Giuseppe Cavuoto*	**47**
Chapter 3	Marine Geophysics of the Naples Bay (Southern Tyrrhenian Sea, Italy): Principles, Applications and Emerging Technologies *Gemma Aiello and Ennio Marsella*	**61**
Chapter 4	Oceanic Oscillation Phenomena: Relationships with Synchronization and Stochastic Resonance *Shinya Shimokawa and Tomonori Matsuura*	**123**
Chapter 5	Assessment of Ocean Variability in the Sicily Channel from a Numerical Three-Dimensional Model Using EOFs Decomposition *R. Sorgente, A. Olita, A. F. Drago, A. Ribotti, L. Fazioli and C. Tedesco*	**137**
Chapter 6	Monitoring Test of Crack Opening in Volcanic Tuff (Coroglio cliff, Italy) Using Distributed Optical Fiber Sensor *Aldo Minardo, Mauro Caccavale, Agnese Coscetta, Giuseppe Esposito, Fabio Matano, Marco Sacchi, Renato Somma, Giovanni Zeni and Luigi Zeni*	**179**
About the Editor		**189**
Index		**191**

PREFACE

This book presents current research in the field of geophysics, particularly referring to some principles, applications and emerging technologies. Topics discussed comprehend environmental geophysics, with an explanation of geophysical methods ("seismic refraction and reflection techniques," "seismic analysis of surface waves," "electrical methods," "gravimetric and microgravimetric methods," "magnetic and electromagnetic methods") and new methodologies of processing and interpretation with examples in the area of Greece; applications of innovative geophysical techniques in other coastal areas, showing a new approach to seismic data acquisition, joining up land and marine seismic data through a marine streamer connected to a land cable with examples in Procida and Ventotene islands (Southern Italy); marine geophysics of Naples Bay focusing on seismic and magnetic data offshore the Vesuvius, Phlegrean Fields, Ischia and Procida volcanic complexes and a technological evolution in seismic data acquisition from explosive, Sparker and Watergun seismic sources towards new techniques, including MEAS (Multispot Extended Array Sparker) used in marine geological mapping of Southern Italy's continental shelf and slope. Other topics in this book include oceanographic aspects regarding the oceanic oscillation phenomena and their relationships with the synchronization and stochastic resonance, particularly referring to the transitions among stable and unstable states of western boundary currents, such as the Kuroshio in Japan and investigating their occurrence in general oceanographic circulation as a rhythmic natural phenomena; oceanographic aspects regarding the assessment of ocean variability in the Sicily Channel (Southern Italy) through a numerical 3D model based on Empirical Orthogonal Functions (EOF) analysis aimed at studying the seasonal and interannual variability of oceanographic circulation and hydrology, and extending down to the African coast, involving the Tunisian and Lybian continental shelves (Mediterranean sea); and finally, experimental application of distributed optical fiber sensors through Brillouin Optical Time Domain Analysis (BOTDA) to the monitoring of artificially induced cracks opening in the volcanic rocky slope of the Coroglio coastal cliff (Naples, Southern Italy) aimed at the development of early warning systems and geomorphological setting of coastal cliffs, characterized by a rocks falls, topples and slides.

Chapter 1 – The rapid development of industrial and urban areas in the last decades caused a significant amount of environmental degradation with serious consequences to human life, ecosystems, natural resources, land utilization, as well as to human and natural heritage. Thus, the role of site restoration and rehabilitation projects, in the context of the Management Environmental degradation, is crucial for the modern society. Environmental

Geophysics comprises a relative modern part of the Geophysics discipline, aimed at determining the physical properties of the shallow sub-surface by using non invasive high-resolution geophysical techniques. The detailed knowledge of the subsurface structure is valuable for a variety of disciplines involved in the study of the shallow earth's crust, that is directly related with human activities. In particular, Environmental Geophysics efficiently deals with technical problems arising during construction works such as karst collapse, evaporite sinkholes, landslides, mine subsidence and fault zones. Furthermore, it is involved in investigations concerning the detection of buried or abandoned waste, the location of the aquifer and part of its related properties to identify the possible presence of contaminants in the subsurface and finally, the mapping of archaeological sites. The aim of this work is to present the most familiar and robust techniques used in environmental geophysics and new methodologies of data processing and interpretation. Extensive literature research and examples from case studies are presented and analyzed. Special emphasis is given to the limitations of all applied techniques.

Chapter 2 – Coastal and transitional areas represent highly dynamic, quickly-changing environments. These areas are subject to multiple interactions with the marine-land-fluvial systems processes. Different types of processes such as coastal erosion, seawater intrusion, pollution and transport of sediment affect these areas. The geophysical surveys permit to improve the understanding of these phenomena and consequently to assess the vulnerability and the potential of the coastal zone. The development of geophysical models integrated into management coastal system is the key to achieve reliable prediction model. Moreover, coastal areas are among the most populated and urbanized areas. For these reasons, the application of geophysical minimally invasive methods, such as seismic and geoelectric techniques, are the most appropriate tool to provide geology and to collect data on land-sea zone. In coastal areas, the seismic reflection/refraction survey plays a key role to increase the knowledge of civil engineers. In particular, seismic survey provide information on subsoil geometry and evaluate the hydraulic and mechanical characteristics necessary in the planning, design and construction stages. The seismic method, routinely employed to obtain information on subsurface geology, has a limited use in the transitional zone. This chapter describes a non standard approach to acquire seismic data in the backshore-foreshore area, joining up land and marine seismic data, using a marine streamer connected to a land cable. This method allows to obtain a seismic site characterization in terms of primary wave velocities and shear wave velocities (both the P-wave and S-wave velocities are used to define the elastic modules and geotechnical parameters), to identify major acoustic impedances within the overburden (e.g., bedrock distribution), to delineate any local subsurface structure (e.g., existence of weak zones such as faulted areas or weathering zone). The monitoring of the intrusion of seawater also represents a priority for the safeguard of coastal areas. The identification of changes in the freshwater-saltwater interface position can be a useful element for the rationalization of water resources in order to guide the choice of use of these areas. To address this issue, we use electrical surveys, in particular 3D Electrical Resistivity Tomography, to monitor and image fresh-salt water transition.

Chapter 3 – Marine geophysics of the Naples Bay is discussed focusing particularly on some principles, applications and emerging technologies in seismic and magnetic marine data. The case histories of Somma-Vesuvius volcanic complex, Phlegrean Fields offshore and Ischia and Procida islands offshore (Naples Bay) are here discussed. Seismo-stratigraphic techniques and methodologies are explained, with the example of the Naples area, where the widespread volcanic activity during Quaternary times has disallowed for the application of a classical stratigraphic approach due to the occurrence of interlayered sedimentary sequences and interstratified volcanic bodies. Principles, applications and emerging technologies of marine geophysical data in the Naples Bay are considered with respect to the data acquisition, processing and interpretation, taking into account some historical aspects towards new technologies. Among the sources tested in the Naples Bay there are the explosives the Sparker and the Watergun while the details to study geomorphological data have been analyzed through the Surfboom and the Sidescan Sonar. MEAS (Multispot Extended Array Sparker) seismic source has been largely used to acquire large database of single channel reflection seismics. More recently, by means of Multitip SAM96, SAM400 transducer high resolution seismic data in the Bay of Naples have been recorded and interpreted in detail. Significant correlations among seismic and magnetic data have been attempted based on comparative analysis of seismic and magnetometric datasets recorded in the Naples Bay.

Chapter 4 – Various oscillation phenomena, such as the transitions between stable and unstable states of western boundary currents, i.e., the Kuroshio, are present in the oceanic circulation of the North Pacific Ocean. These phenomena are considered to be related to non-linear rhythmic phenomena, such as the synchronization and the stochastic resonance, which are often observed in non-linear ocean systems. Synchronization is an adjustment (e.g., frequency and/or phase locking) of the rhythms of two or more self-sustained oscillating systems that have different periods because of their non-linear interaction. Stochastic resonance is a phenomenon in which parts of potential signals can exceed the threshold when adequate noise is added to non-linear systems; these can then be detected as actual signals. However, oceanic applications of these non-linear rhythmic phenomena have not been investigated in detail. Thus, we investigated the responses of oceanic double gyres to external wind forcing with and without noise using a 1.5 layer quasi-geostrophic model and considered the possibility of the occurrence of these non-linear rhythmic phenomena in general oceanic circulation.

Chapter 5 – The circulation in the Sicily Channel and the surrounding areas has been simulated from January 2001 to December 2004 using a numerical three-dimensional, free surface, nested limited area ocean model. Basic monthly-based statistics, as well as Empirical Orthogonal Functions (EOF) analysis, have been used to study the seasonal and interannual variability of the circulation and hydrology of the Sicily Channel. Significant interannual variability of the surface properties, superimposed over an underlying and stronger seasonal cycle, has been identified in the space-time domain through the EOFs and has been correlated with the changes in the simulated circulation and hydrology. The main interannual event is related to changes in the surface circulation that occurred during summer 2003. The skill of the numerical model is assessed to measure its ability in reproducing the dynamical characteristics of this part of the Mediterranean, taking into account the general lack of data, especially in the southern part of the domain. The model performances also suggest directions for further research efforts and for model improvements, such as an improved reproduction of the mixed layer.

Chapter 6 – The authors report the experimental application of distributed optical fiber sensors based on stimulated Brillouin scattering (SBS) through the so-called Brillouin Optical Time Domain Analysis (BOTDA) to the monitoring of artificially induced crack opening in a volcanic rock slope. The aim of this chapter is to show how the sensing optical fiber cable is able to detect the formation and to follow the growth of fractures in the tuff rocks and to identify their location along the cliff. The experiments have been performed at the base of the Coroglio tuff cliff in Naples (Italy) and have demonstrated that early detection of crack opening can be obtained and development of early warning systems is an attainable goal of the research.

In: Geophysics
Editor: Gemma Aiello

ISBN: 978-1-63484-831-2

Chapter 1

ENVIRONMENTAL GEOPHYSICS: TECHNIQUES, ADVANTAGES AND LIMITATIONS

Pantelis Soupios[1] and Eleni Kokinou[1,2]
[1]Department of Environmental and Natural Resources Engineering,
Technological Educational Institute of Crete, Heraklion, Greece
[2]Dynamics of the Ocean Floor,
Helmholtz Centre for Ocean Research Kiel, Geomar, Germany

ABSTRACT

The rapid development of industrial and urban areas in the last decades caused a significant amount of environmental degradation with serious consequences to human life, ecosystems, natural resources, land utilization, as well as to human and natural heritage. Thus, the role of site restoration and rehabilitation projects, in the context of the Management Environmental degradation, is crucial for the modern society.

Environmental Geophysics comprises a relative modern part of the Geophysics discipline, aimed at determining the physical properties of the shallow sub-surface by using non invasive high-resolution geophysical techniques. The detailed knowledge of the subsurface structure is valuable for a variety of disciplines involved in the study of the shallow earth's crust, which is directly related with human activities. In particular, Environmental Geophysics efficiently deals with technical problems arising during construction works such as karst collapse, evaporite sinkholes, landslides, mine subsidence and fault zones. Furthermore, it is involved in investigations concerning the detection of buried or abandoned waste, the location of the aquifer and part of its related properties to identify the possible presence of contaminants in the subsurface and finally, the mapping of archaeological sites.

The aim of this work is to present the most familiar and robust techniques used in environmental geophysics and new methodologies of data processing and interpretation. Extensive literature research and examples from case studies are presented and analyzed. Special emphasis is given to the limitations of all applied techniques.

Keywords: applied geophysics, environmental and engineering geophysical methods

1. INTRODUCTION

Environmental Geophysics [1, 2, 3] deals with the estimation of physical properties in the shallow subsurface. The characterization of the near-surface structure is implemented by applying a variety of geophysical methods, i.e., active and passive methods. Active methods, such as seismics (reflection/refraction), electric/electromagnetic and radar techniques, use artificially generated fields on the earth's surface to further measure and interpret the earth's response. On the other hand, passive techniques [4] such as magnetics, self potential and gravity, measure, analyze and interpret signals, supplied by the natural fields of the earth. The ultimate goal of the above mentioned geophysical techniques is to collect all the information regarding the near surface interior and to further contribute to the construction and protection of the human infrastructures, the protection of the life, health and heritage of living beings and the conservation of the natural resources. More specifically, environmental geophysics is involved in the investigation, detection and monitoring of:

- Construction works [5, 6, 7, 8, 9, 10] such as highways and railways, dams, bridges, abandoned pipelines and tanks, buried and unexploded ordnance-UXO and any related technical problem.
- Dangerous contaminants to public health and the environment [11, 12, 13, 14, 15, 16, 17, 18, 19, 20 and 21] such as LNAPLs, DNAPLs, heavy metals, fertilizers, in the soils and the groundwater.
- Geological hazards [22, 23, 24, 25, 26, 27, 28, 29, 30 and 31], such as sinkholes, landslides, karst collapse, faults and deformed stratigraphic units.
- Mapping of archaeological sites [32, 33, 34, 35, 36, 37, 38, 39 and 40].

Environmental Geophysics generally applies non-destructive techniques to investigate the shallow subsurface in limited and wide areas, under certain conditions: 1. there are contrasts in the physical properties of the ground, 2. during the acquisition process, the optimal sampling parameters (in space and/or time) are selected, taking into account all available information corresponding to the ground conditions, 3. more than two geophysical techniques are applied in order to constrain the final solution and evaluate the results. 4. proper data processing, that comprises a crucial step in a geophysical project, as the selection of the wrong processing steps can lead to completely distorted results and further misinterpretation.

In the context of this work, we present a comprehensive approach of the most familiar and robust techniques in Environmental Geophysics, based on literature research and the analysis of the results and the problems arising in case studies, bearing in mind that often there is no single optimum technique, applicable to each case.

2. ROBUST TECHNIQUES OF ENVIRONMENTAL GEOPHYSICS

Well tested methods in Environmental Geophysics are seismic (surficial or crosshole passive or active seismic methods), electrical (1D, 2D, 3D or 4D electrical resistivity imaging techniques, time domain or spectral induced polarization methods, self potential) electromagnetic (including terrain conductivity, ground penetration, transient electromagnetic measurements), gravity-microgravity and magnetic (including magnetic susceptibility). In the present paragraph, the principles, the survey planning, the data acquisition, the interpretation and the limitations of the aforementioned methods are presented. Generally, when we plan a geophysical investigation, we should always bear in mind:

- To collect any information (geological, remote sensing, topographic, engineering, geophysical, etc.) concerning the study area.
- To know exactly the cost, the time of the survey and the size of the area.
- To estimate the possible noise sources in the local environment, affecting the applied geophysical methods, in order to decide the most effective combination. Noise tests should always be carried out prior to data acquisition.
- To estimate the detectability of the target and make statistical analyses concerning the expected range of the anomaly.
- To decide the expected lateral and vertical size of the target in order to select the appropriate sampling interval (ideally, at least 5 data points within the half width of the anomaly [3] and the station/grid spacing.
- Forward modeling concerning the expected amplitude anomaly of the target [41, 42, 43].

2.1. Seismic Methods

2.1.1. Seismic Refraction Technique - SRefr

Seismic techniques can successfully reconstruct, with high resolution, the lateral and depth variation of seismic velocity, producing detailed images of the subsurface. The method is based on the fact that different geomaterials have different rock properties such as density, porosity, permeability and compaction, which can be recognized/recorded during seismic measurements. On the other hand, the amount of data usually collected during a routine fieldwork can rapidly become overwhelming and the logistics of data acquisition are more intense and expensive than in the other geophysical methods. In general, the equipment used for acquisition is more expensive than other instrumentation and data processing and interpretation can be time consuming.

In addition to the aforementioned limitations, there is one more inherent limitation related to the development of the method. The seismic method was originally developed by the oil industry and it was mainly intended to image the subsurface at depths of hundreds or thousands of meters. As a result, conventional seismic acquisition systems use very energetic sources and a large number of geophones. The use of seismic methods for imaging very shallow targets is a relatively recent development [44, 45], and consequently its application in near surface experiments is still in progress. Shallow seismic experiments use portable seismic acquisition systems which have a limited number of channels (24, 48 or 60 is a typical number), single geophones and relatively weak (and cheap) surface sources (usually a sledgehammer). In this case, a survey of this scale can be performed very quickly by at least two people, so that acquisition costs and time will be similar to those for a GPR survey.

To define the applicability of the seismic techniques in shallow application such as the environmental and engineering projects, one should consider which physical properties of the studied targets will make them detectable. In general, any velocity anomaly which is considered as a signal/target of interest can be defined in the shallower part (2-100m) of the earth. In some engineering applications, the depth of investigations can be higher depending on the structure and its foundation. In such applications, the seismic method can detect bedrock discontinuities, possible fracture zones, stratigraphy and water saturation which can be connected with liquefaction phenomena.

Originally, seismic surveys had sources and receivers located only on the earth's surface, and by collecting the data along a line, a seismic model was expected to be estimated applying refraction or reflection tomography. More recently, different kinds of data/information (seismic surface waves) acquisition have been developed to reconstruct the 2D/3D S-wave model of the study area. Seismic data can be collected into a borehole or a pair of boreholes deploying geophones and sources on the surface or into the boreholes, increasing the resolution of the resulted seismic model. Finally, depending on the size or the purpose of the project, the scale of the experiment can be changed, collecting and processing data in meso- and/or micro-scale.

The seismic refraction method can be applied for stratigraphic mapping, estimation of depth of the bedrock and the water table, prediction of the rippability of specific rock types, locating sinkholes, landfill investigations and geotechnical investigations.

2.1.2. Seismic Analysis of Surface Waves - SASW

Surface waves allow the measurement of the variation of S-waves in soils which can be transformed to stiffness with depth. Surface waves (Rayleigh and/or Love waves) have the property that ground motion becomes negligible below a depth of one wavelength. By recording surface waves of different frequencies (producing the dispersion curve), and therefore wavelengths, the properties of the subsurface can be determined at different depths. The characteristic velocity of the surface waves can be determined by measuring the signals received at a series of geophones. Once the shear wave velocity profiles are determined, shear (G_{max}) and Young's moduli of the materials can be calculated through the use of simple mathematical equations. These geotechnical parameters can be used by civil engineers for managing the foundation of large scale structures (bridges, dams, wind turbines, etc.). A stiffness profile can be generated for each adjacent pair of geophones. Profiles can be combined to produce cross-sectional images of the properties of the subsurface. Similarly, a number of adjacent cross-sections can be combined to produce a pseudo-3D image.

This method has been mainly used for engineering purposes, such as stratigraphic mapping prior any foundation, estimation of the depth of bedrock including landfill base location, soil stiffness and ground improvement verification and in-situ ground stiffness for large-scale geotechnical and civil engineering projects.

2.1.3. Seismic Reflection Technique - SRefl

The seismic reflection method has the highest resolution among the other seismic methods and identifies variations in material type with depth and horizontal position. The method was sponsored by the oil industry to provide detailed 2D and 3D images of the subsurface to the depth of thousand of meters. This technique images the interfaces between geological formations with contrasting acoustic or elastic or density properties and, consequently, contrasting seismic velocities. Mapping of these contrasts across an area can identify the extent and depth of specific layers or interfaces of interest (specific tectonic/stratigraphic structures which usually host oil called oil traps).

The SRefl method is based on the propagation of seismic waves through the subsurface and on their reflection at interfaces across, which there is contrast in velocity and/or other acoustic properties. In general, seismic energy is generated and the seismic waves travel through the subsurface. They are recorded by geophones at different positions along the survey line. Each geophone is connected to the seismograph (the acquisition system) which records the arrival time and magnitude of the induced voltages (oscillations) at each geophone.

Seismic reflection surveys are mainly used to map subsurface geological boundaries and stratigraphic variations. A key advantage of the technique is that, after detailed data processing, it can provide an accurate cross-section image of the subsurface (Figure 1).

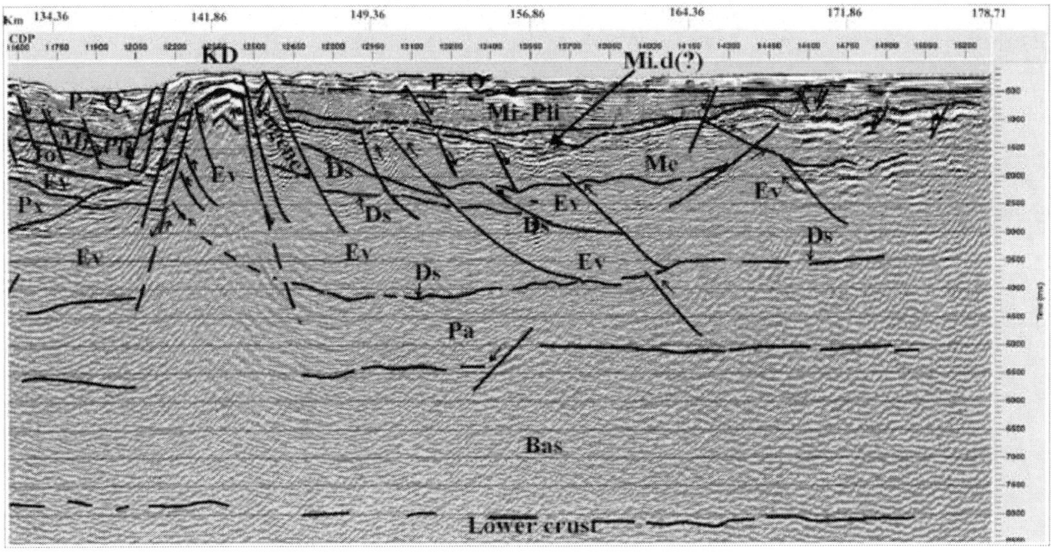

Figure 1. Interpreted deep seismic reflection section (migrated) from ION-7 (modified after [46]).

2.1.4. Advantages and Limitations of Seismic Methods

The conventional *refraction seismic method* has the following advantages and limitations:

Advantages: Refraction measurements generally use fewer source and receiver pairs for producing a reliable subsurface model and thus, the method is relatively fast and cheap to be applied. A few processing steps are needed on raw seismic data to estimate the arrival times of the first breaks on recorded ground motion (seismograms).

Disadvantages: Refraction only works if the velocity increases with depth (this problem was solved by using tomographic approaches solving the forward and inverse problem). The observations are generally interpreted in terms of layers-horizons. This method uses only the travel times between sources and receivers, excluding any other information that the waveforms have, such as the amplitude. Thus, the reconstructed velocity model is usually quite poor and has limited resolution.

The *reflection seismic method* has the following advantages and limitations.

Advantages: Reflection measurements are collected at small source-receiver offsets to ensure near vertical ray paths. Processing and interpretation of the reflection seismic method are independent from the velocity variation with depth. Reflection seismics are able to reconstruct and depict complex geological models and make use of the whole waveforms.

Disadvantages: Since many source/receiver pairs must be used to produce detailed models of the subsurface, the method is very expensive. Reflection seismic processing can be very computer consuming, requiring huge CPU time for calculations and sophisticated software for processing by high - level expertise personnel. Thus, the processing of reflection seismic observations is relatively expensive.

The seismic analysis of surface waves has the following advantages and limitations.

Advantages: The SASW/MASW (MASW - multichannel analysis of surface waves) technique is the optimum seismic technique for predicting the modulus of base materials in order to have a reliable geotechnical characterization of the study area. Due to multichannel recording (many acquired data) and processing schemes employed (many automated commercial software are available), results (Vs information) of the survey are highly accurate, even under the presence of higher modes of surface waves or cultural noise. For the same reason, the processing steps can be fully automated. Thus, the method is extremely user-friendly, easy and fast to implement.

Disadvantages: Due to the inner properties of surface waves, the resolution is highly dependent on the size of the anomaly. An empirical rule says that the minimum size that can be resolved is about one tenth of depth.

2.2. Electrical Methods

2.2.1. Electrical Resistivity Imaging - ERI

The electrical properties of the subsurface change due to the variation of the near surface earth material, with the presence or not of fluids that saturate the subsurface and to the occurrence of buried objects representing the target. Electrical techniques attempt to reconstruct the subsurface based on changes of the aforementioned properties distribution as a function of depth and horizontal distance. The most commonly used electrical technique is the Electrical Resistivity Imaging (or Electrical Resistivity Tomography, ERT) which can be applied in 2D, 3D and 4D (time lapse) mode. Simplifying the acquisition of ERI method, measurements of ground resistance are collected by introducing an electric current pulse into the subsurface using two metal electrodes (called current electrodes), planted onto the ground along a profile or into a borehole to collect well measurements. The current, passing through the subsurface, produces changes in electrical voltage (between two potential electrodes) in several places of the subsurface depending on the acquisition configuration. Applying the Ohm's law, the measured voltage can be "translated" to resistance reading for the subsurface between the two potential electrodes.

The ERI method can be used for landfill investigations, mapping and monitoring contaminated plume, groundwater pollution and infiltration studies, determination of the depth of the bedrock and location of sinkholes, buried channels, dykes, ore bodies, fissures, faults and mineshafts.

2.2.2. Induced Polarization Imaging - IPI

Induced polarization (IP) imaging is a complementary method to ERI and deals with the capacitance of the subsoil. The subsurface dissipates and stores the electric current flowing through it. The resistivity methods measure the resistance of the subsurface to energy flowing, whereas IPI method measures the amount of stored energy (referred to as chargeability). In theory, when a current flows through the subsoil, a small amount of energy is stored at the soil structure (minerals) and the soil becomes charged. When the current is turned off, this charge decays with time and can be recorded using potential electrodes. The estimation of the rate of this decay with time can provide the chargeability of the formation. Very often, IPI method can be very accurate since two formations can have similar resistivity but different charge abilities. This is the main reason that IP imaging is suggested as complementary method to ERI since it can provide additional information of the subsurface materials. The IPI method has the same applications as the ERI method.

2.2.3. Self Potential method - SP

Self potentials (SP) are the difference in natural electrical potentials between two measuring points. These potentials can be produced by groundwater flow, contaminants movement, mineral deposits and chemical diffusion (electrokinetic phenomena). The magnitude (less than a millivolt to over one volt) and the polarity of self potentials are used for the interpretation of SP anomalies. The most common factor for SP anomalies is groundwater, and is generated by the flow of conductive water through the resistive medium, or by the involvement of water in natural chemical reactions.

The SP method is mainly applied in hydrogeological investigations, landfill delineation, geothermal surveys, identification of ore bodies and detection of leaks in reservoirs and dams.

2.2.4. Advantages and Limitations of Electrical Methods

Advantages of the electrical methods are: The equipment is portable and inexpensive, compared with the instruments used in other geophysical methods. The user can have a qualitative interpretation of the collected data in a very short time quickly after the completion of the field measurement. Concerning the field crew, three experienced persons are enough to collect about 5-7 profiles per day depending on the total length of the profile and on the morphology of the study area. The method can be used for shallow or deep surveys, since the penetration depth depends on the electrode spacing and on the power of the instrument (in some cases an external booster is needed).

Disadvantages of the technique are: In cases of deep surveys (for mineral or geothermal explorations) larger electrodes spacing (about 20-50 meters) are required. Such long cables are heavy and consume much field time and personnel. The interpretation of the study area is not always straight forward since the geological and tectonic structure is complex. In such cases, a verification of the resulted tomographic images is needed using available borehole logs or other information which can constrain the resulted model.

Some of the *advantages of IP data* are: Having complementary IP data can help the interpreter to remove the ambiguity and reconstruct a better and more reliable subsurface model. The reason is that the IP measurements can distinguish the conductive clay from a conductive brackish saturated sand or gravel. The IP measurements are valuable to archaeological prospection and mineral exploration.

2.3. Gravity-Microgravity

The gravity method was initially used in oil, gas and mining exploration and later in urban planning and engineering for detecting possible subsurface technical problems during construction works, in archaeology and in hydrologic investigations. A great deal of work [47, 48, 49, 50, 51, 52, 53, 54, 55, 56, 57, 58, 59 and 60] has been done in the field of gravity, since it belongs to the oldest geophysical techniques. The gravity technique determines the earth's gravitational attraction, that is directly proportional to the mass and consequently to the density of the subsurface materials. By using the gravity method in geophysics, we are trying to specify density contrasts. A density contrast represents a gravity anomaly (Figures 2, 3), corresponding to a local deviation of the background gravity field due to the presence of a subsurface geological feature or human construction (i.e., graves, tunnels etc.) with specific shape and geometry. The geological feature, causing the positive or negative gravity anomaly, could be a stratigraphic discontinuity or contact, a fault or a fold, a diapirism and an igneous intrusion. The units, commonly used in geophysics for the gravity anomaly are 1 mGal = 10^{-3}Gal, 1 gravity unit (gu) = 10^{6} ms^{-2} = 1 μms^{-2}, 1 gu = 0.1 mGal = 100 μGal.

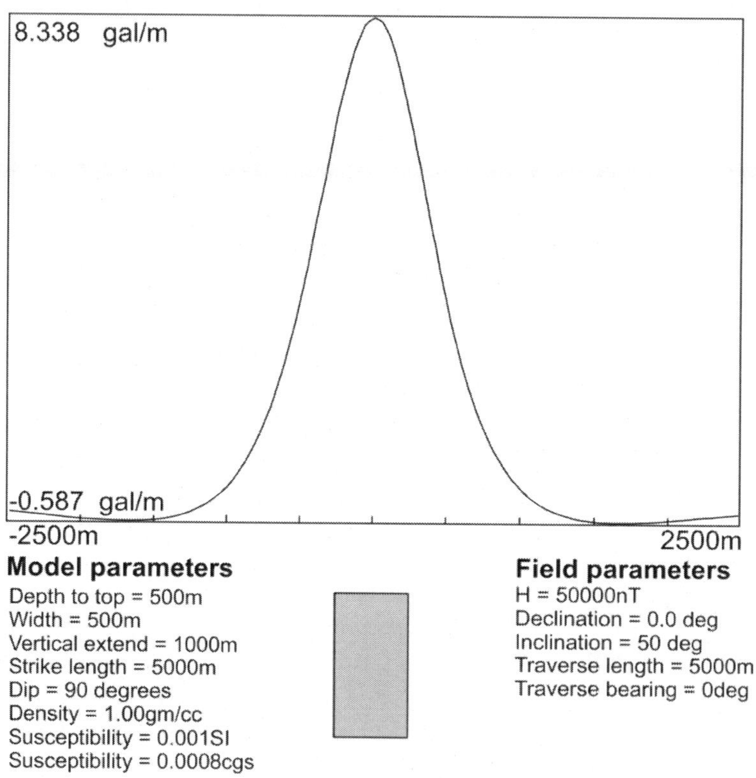

Figure 2. Vertical gradient of the gravity corresponding to a prism. Modeling has been done by the free software pdyke [61] by Geophysical Software Solutions.

2.3.1. Advantages and Limitations of the Gravity-Microgravity Method

The gravity-microgravity method is very effective in detecting targets that present a considerable density deference compared with the surrounding geological formations and targets underlain high velocity zones [63]. Remote and difficult to access areas can be mapped using airborne gravity or marine gravity. The use of inexpensive workstations and personal computers has generally reduced the cost of this method (acquisition, processing and interpretation), while at the same time the quality and accuracy of the results has increased due to the rapid progress in software industry and the use of global positioning system (GPS).

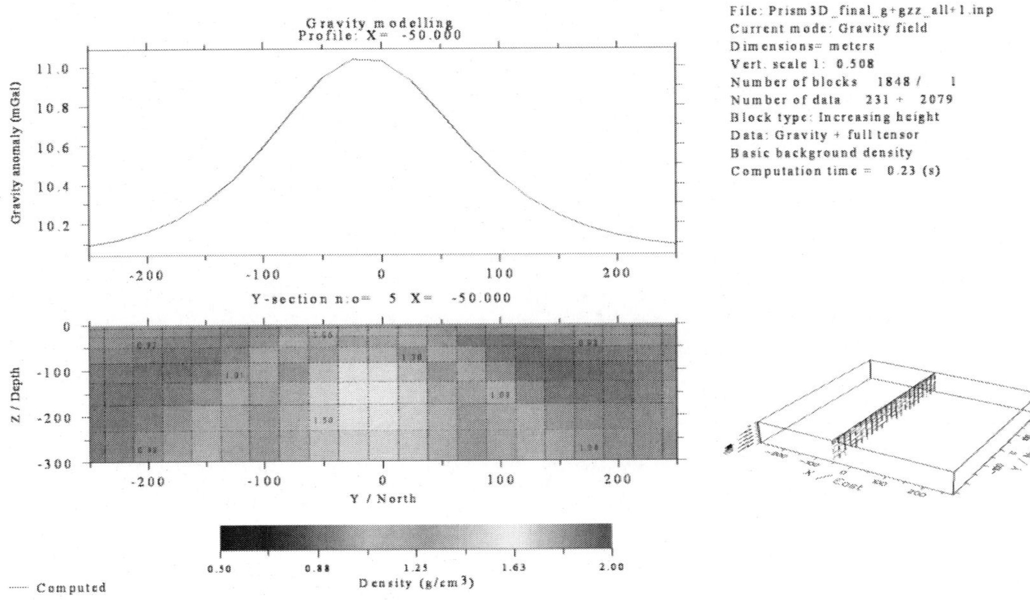

Figure 3. Vertical gradient of the gravity corresponding to a 3D prism. Modeling has been done by the free software Grablox2 [62].

The *limitations of the gravity/microgravity method* are:

- The gravimeter has to be placed very carefully on level ground.
- After every reading the instrument must be locked to avoid drift due to excessive vibration.
- The near and far topographic anomalies around the station, across the profile or in the grid, must be accurately mapped.
- Ground vibrations during the data acquisition strongly affect the accuracy of the value taken.
- The calibration of the instrument often needs more than one control points, that may be time consuming.

2.4. Magnetic Survey

Magnetic prospecting comprises a few centuries-old detection method, first applied in iron ore exploration. These days, the magnetic method is a fast and cheap method to acquire geophysical data for engineering and environmental purposes such as the detection of pipes, cables, tanks, drums, piles, metal objects and toxic waste, archaeological relics and old fire pits, geological features (magnetic basement, faults and folds) and finally for the analysis of tectonic processes.

Magnetic measurements are used to estimate small, localized variations in the Earth's magnetic field.

In Environmental Geophysics, the magnetic unit used is the 1nT (1 billionth of a Tesla) or the 1gamma = 1nT. The magnetic response of a buried metal object may be a few hundreds of nT while archaeological relics (walls), in weakly magnetic sediments, correspond up to a few dozen of nT.

Although both magnetic and gravity methods belong to passive techniques, measuring the corresponding earth's potential field have some distinct differences between them: 1. The gravity field is due to one or more monopoles, while the magnetic field is always produced by a dipole, 2. the magnetic field is less stable than the gravity field, in terms that it is more sensitive to spatial and temporal changes.

Magnetic Susceptibility

Rock-magnetic parameters [64] can be used for studying the environmental systems, the ground and marine soil magnetism, the fluvial processes, the lake sediments, the sources of magnetism in atmosphere, the palaeomagnetism and the biomagnetism. In particular, environmental studies [65, 66, 67, 17, 68, 18] based on magnetic properties attract the scientific interest in the last 30 years, focusing on the detection of contamination in sediments and soils in drainage systems, topsoils and vegetation, urban soils and ground pollution due to industrial emissions. The magnetic parameters of soils and rocks comprise a rapid and cheap indicator of industrially polluted areas and help in the targeted selection of samples for further geochemical analysis.

Magnetic susceptibility (χ, as the ratio of the temporary magnetization to the applied field) is the most often used parameter in every environmental magnetism survey. More specifically, the magnetic susceptibility is a measure of how easy a material can be magnetized [64] and is defined by the quantities of volume susceptibility k (dimensionless) and mass specific susceptibility χ. Units of mass specific susceptibility of 10^{-6} m^3kg^{-1} are commonly used in Environmental Geophysics. Accurate measurements of mass susceptibility are usually obtained in two frequencies (f_{low} = 0.43KHz and f_{high} = 4.3KHz). The magnetic behavior of many minerals changes with temperature. Curie temperature is the temperature point above which all magnetic minerals become paramagnetic. So, based on the shape of the temperature - susceptibility curves, the presence of magnetic minerals and domains can be detected. Anhysteretic remanent magnetization (ARM) and isothermal remanent magnetization (IRM) are often used in environmental magnetic investigations. Both properties are estimated by exposing a sample to an external magnetic field.

2.4.1. Advantages and Limitations of the Magnetic Method

The magnetic method is very effective in detecting low depth targets with high accuracy such as archaeological relics, geological features related to heat sources, intrusion of volcanic bodies, the aquifer and any kind of metal objects. The data acquisition and processing is fast and of low cost. Additionally, this method also works effectively in areas with irregular morphology. The *limitations of the magnetic prospecting* in Environmental Geophysics are:

- The survey area must be free of steel, iron metals and cables, because they influence the data acquisition. If an iron fence is present, measurements have to be taken more than 7-10m away.
- The magnetometer operator has to be free from any metal objects containing steel or iron. If the operator takes readings around the sensor and they do not change by more than 1 or 2 nT, data acquisition can start.
- The nature of the geological formation on ground comprises an important parameter affecting the measurements. For example "terra rossa," that is by its nature rich in ferrous minerals, generally causes noise during the data acquisition.
- Areas with dense vegetation and very irregular topography are generally more difficult to be surveyed.
- The magnetometer calibration is a very important procedure during a magnetic survey to improve the accuracy of target detection and localization. The instrument has to be calibrated following exactly the instructions of its manual, in the beginning of the survey and prior starting the measurements in the next grid, if many grids are to be surveyed. In addition, in case of sudden temperature changes the instrument has to be calibrated more often.
- Magnetic measurements must stop in case of strong magnetic disturbances of the earth's magnetic field.
- Deep targets are more difficult to detect and produce very weak magnetic responses.

The *limitations of the magnetic susceptibility method* in environmental geophysics are:

- Magnetic susceptibility comprises an indicator of possible heavy metals presence. The results of the pre-mentioned method should always be confirmed by geochemical analyses.
- Subsurface geological structure must be undisturbed in the surveyed area.
- The contact with metal objects has to be avoided during sampling.
- Magnetic susceptibility measurements on site must be taken away from metal objects.

2.5. Electromagnetic Methods

2.5.1. EM Terrain Conductivity (TCM) Method

In near surface electromagnetic (EM) surveys, the measured conductivity is presented as a function of depth along profiles or as a spatial variation. In general, the geological formations and the buried objects (geophysical targets) have different conductivities. By acquiring spatial (in X, Y) conductivity data anomalous areas can be identified. The TC method is based on the induction of electric currents into the ground by the magnetic component of electromagnetic waves generated at the surface. An alternating current, of variable frequency, is passed through a coil of wire (a transmitter coil). This process generates an alternating primary magnetic field which, in turn, induces very small eddy currents in the earth, the magnitude of which is directly proportional to the ground conductivity in the vicinity of the coil. These eddy currents then generate a secondary magnetic field, a part of which is intercepted by a receiver coil. The interaction between the primary and secondary magnetic flux and the receiver coil generates a voltage that is related to the electrical conductivity of the subsurface, expressed as milliSiemen/meter (mS/m).

2.5.2. Time Domain Electromagnetic Method - TDEM

The TDEM is a new method applied in geophysics over the last two decades. There are several textbooks presenting the method [69, 70, 71, 72, 73, 74, 75, 1, 76 and 77].

The TDEM is a controlled source electromagnetic (CSEM) method for shallow (archaeological and groundwater) [40, 76, 78] and deeper (geothermal, mineral, oil and gas) applications. Every CSEM system consists of a transmitter and a receiver, Tx and Rx dipoles (loops) respectively, with varying sizes depending on the required penetration depth. In general, an increase in transmitter loop increases the signal to noise ratio (S/N) resulting in an increase of the exploration depth [79].

In theory, a current flowing through the transmitter loop, generates a primary field. Applying an on-off switching sequence, a transient field is generated and is measured by the receiver loop. The decay rate of the electromagnetic field depends on the resistivity structure of the subsurface. Thus, the measured voltage on Rx coil can provide information about geoelectrical structures at several depths [80].

2.5.3. Ground Penetrating Radar

The Ground Penetrating Radar (GPR, [81]) method is a high-frequency electromagnetic technique commonly applied to solve engineering and environmental problems [32, 82, 83, 84, 85, 86] such as mapping of the shallow geological subsurface structure (stratigraphy and disturbances, water tables, voids and cavities), locating metallic and non metallic manmade structures (pipes, tanks, cables, landfill boundaries etc.) and characterizing archaeological sites.

The GPR technique partly works like the seismic reflection method, considering that source and receiver are located at the same point. The high frequency electromagnetic waves, produced by the source are transmitted in the Earth's interior and reflected on discontinuities dividing layers with different electrical properties (conductivity and the dielectric constant). The remaining part of the electromagnetic energy is diffused at deeper levels. The travel time of the electromagnetic wave from the source to a discontinuity and back to the receiver is recorded and if the propagation velocity of the electromagnetic wave is known then the depth of the reflector can be determined. Electromagnetic waves travel at the same speed in a vacuum/air, i.e., the speed of light $3x10^8$ ms^{-1}. Obviously, the speed of the electromagnetic waves travelling in the earth interior is much lower than $3x10^8$ ms^{-1}. The absorption of the electromagnetic waves is responsible for limiting the maximum penetration depth of a GPR. As this absorption is greater for higher frequencies, low frequency electromagnetic waves are able to propagate deeper in the earth. For instance, a GPR system working in the range of 25-50 MHz can investigate depths of over 50 meters in soils with low conductivity (smaller than 1mS/m) like sand and gravels. On the other hand, the spatial resolution of the method also decreases as the frequency is decreased.

In a profiling survey the GPR system consists of a transmitter and a receiver which are towed along transects. Radargrams are formed through the registration of the amplitude of the reflected signals and represent the time difference between the transmitted and reflected signals. Knowing the velocity of propagation of the EM waves over specific soils, it is possible to obtain a depth estimate of the suspected targets.

2.5.4. Advantages and Limitations of the Electromagnetic Methods

Terrain conductivity EM systems have the following advantages. For a daily survey, two people are needed. The first one carries a small Tx coil, while the second carries the Rx coil. TC meter is a very flexible and easy to use method, providing an average conductivity of the ground without any contact with the ground (as is required with DC resistivity techniques). The main *disadvantage* is the limited penetration depth (always depending on the intercoil spacings).

The *advantages of TDEM geoelectric sounding* over conventional DC resistivity sounding (VES) are significant. They include the following (Wightman et al. 2003): Surveys over large areas (from land, airborne, seaborne or into boreholes) with high lateral resolution. No contact with the ground is required. Improved resolution of conductive electrical equivalence and with little problem injecting current into a resistive surface layer. The *limitations of TDEM techniques* are listed as follows: It is not the optimum method for very resistive material. The penetration depth depends on used frequency and on Tx-Rx separation.

The *advantages of the GPR system* are: It can produce a detailed map of the vertical stratigraphy within a depth range of 1.5–10 m, depending on the operation frequency and the electrical properties of the ground. Features of sharp discontinuities and dielectric contrasts, such as walls, cellars, cavities, tombs and compacted earth can be quite easily resolved. On the one hand, radar can be easily used in urban areas; on the other hand, it exhibits serious problems in areas of conductive clayey soils because of the strong attenuation of the signal. Amplitude averaging techniques over specific depth ranges of radargrams obtained along parallel GPR transects provide depth (or time) slices of the subsurface and a three-dimensional (3D) reconstruction of it.

The *limitations of the GPR. in environmental geophysics* are:

- The detectability of the instrument depends on the sensor's characteristics and on the target's dimensions/geometry.
- Metal is not penetrated by the radar waves and therefore objects behind a metallic boundary may be obscured and thus are not observable.
- Subsurface geology plays an important role in the depth penetration of the instrument. For example, sand comprises a formation easy to be penetrated by the electromagnetic waves while the clay not.
- It is usually recommended to survey almost flat areas, with no irregularities (no vegetation, rocks, holes etc.). However, it can also be effective in irregular ground, only the data acquisition is less efficient.

3. Data Acquisition, Processing and Interpretation

Forward [87, 88, 89, 90] and inverse modeling [91, 92, 93, 94, 95] comprise the basic tools of data processing [96, 97] and further reliable interpretation [98, 99, 100] in applied sciences, including Environmental Geophysics. Forward modeling is the procedure of synthetic data calculation, considering a known model of the earth's structure. On the other hand, the inversion is the procedure of estimating the Earth's model, often with constraints, using the instrumental data, provided by the survey.

Generally, the data processing flow of the geophysical data, and not only, is an iterative procedure and includes:

- Estimation of the error concerning the field observations, such as noise, location inaccuracies and any problems arising during data acquisition.
- Forward modeling prior the data acquisition, taking into account previous information, concerning the survey area. This step is crucial for the evaluation of the field data and to decide the flow diagram of data processing.
- Discretization of the earth's model, taking into account the previous information and the parameters of the data acquisition.
- Data fitting is the procedure showing how the predicted data (resulted from the forward modeling) matches the instrumental data. The well known least squares (L2 norm) and sometimes L1 norm (corresponding to the absolute differences between observed and calculated data) belong to the most wide used methods of data fitting.
- Inversion performance and evaluation of the results.
- Interpretation of the results.

3.1. Seismic Data Acquisition Processing and Interpretation

Seismic data (2D & 3D) can be acquired either on land or in marine environment. The field crew should have the ability to work on versatile field equipment, acquiring efficient, accurate, reliable and cost-effective 2D and 3D single and multi-component seismic data at any terrain and weather conditions. Of course, special attention should be taken to avoid erroneous measurements due to "strange" terrain and weather conditions. To gain time in processing and provide some preliminary results-models to clients, the field crew should offer a broad suite of in-field seismic data processing services. Based on the experience of the field managers, the exploration risk and cost can be reduced helping the clients receive satisfactory results sooner. Specifically, an accurate survey design can provide the highest quality and accuracy of data with minimum cost. Prior to any experiment in the field, the acquisition designers should estimate and provide the appropriate acquisition and parameterization parameters according to different field conditions to have the optimum imaging of the subsurface.

The operators should have specific knowledge and experience on, 2D and 3D (or even 4D) seismic survey design in different land conditions. VSP survey is designed especially for oil exploration and optimum hardware and equipment selection according to facilities and targets. The optimum equipment and acquisition parameters includes azimuth and offset distribution, geometry determination (Swath, Orthogonal, Brick, Patch, Radial, etc.) and the strategy of energy sources.

Seismic data processing (Figure 4) is an important step before data interpretation and presentation of the final results. Usually, processing incorporates a wide variety of processing routines, such as: standard processing routines, velocity modeling, pre-stack depth/time migration and seismic imaging, ray tracing and tomography, multiples attenuation or elimination, seismic data reconstruction, data enhancement and multi-component data processing. Seismic data transcription, seismic data compression, seismic map digitizing and trace reconstruction are other complementary services.

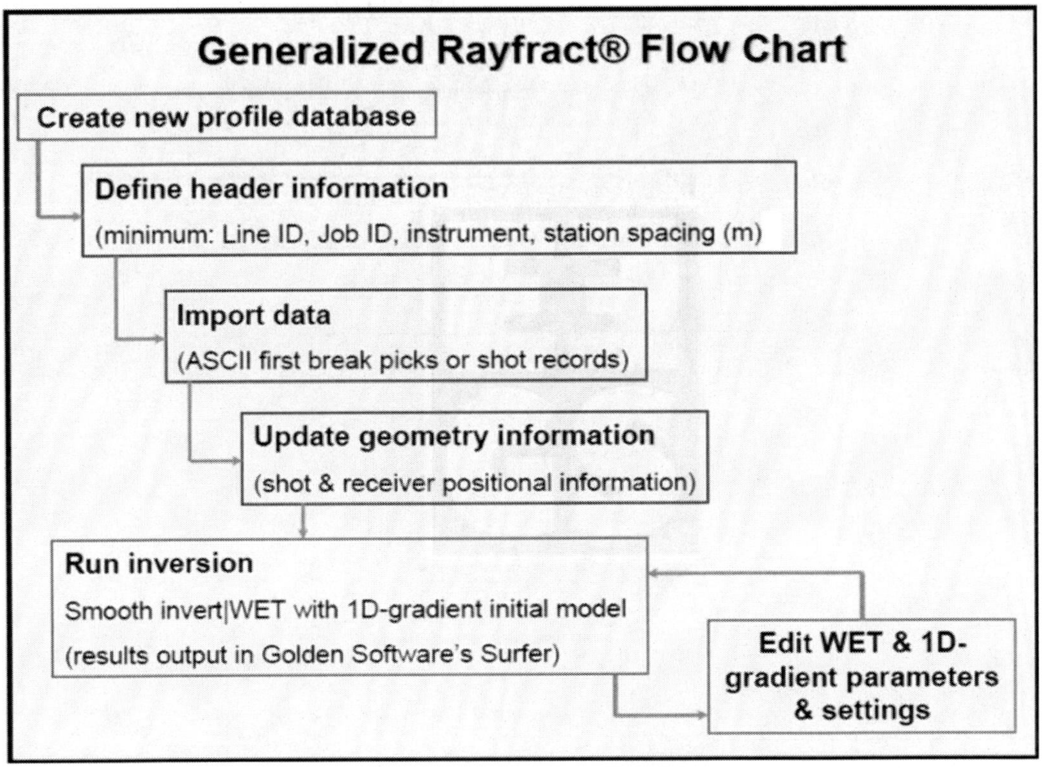

Figure 4. Flow chart of seismic data processing [101].

Interpretation is the end of a very difficult way to knowledge. Months of preparation for fieldwork, data collection and processing are summarized in the final step of data interpretation. Having the appropriate knowledge, experience and the proper equipment to be efficient, enables us to perform seismic data interpretation. Since the interpretation needs a multidisciplinary approach, different kinds of available data like geological data, well logs, gravity and other geophysical data can be used to enhance the final model (Figure 5).

3.2. Geoelectrical Data Acquisition, Processing and Interpretation

Acquisition of geoelectrical data. To construct an ERI cross-sectional image of ground resistance, a multi-electrode cable is connected with electrodes installed on the ground along a straight line (or a crooked line where the location of each electrode should be known to apply the appropriate corrections) with an inter-electrode spacing of 1-50 meters, depending on the requested depth of penetration. Once a measurement of ground resistance has been

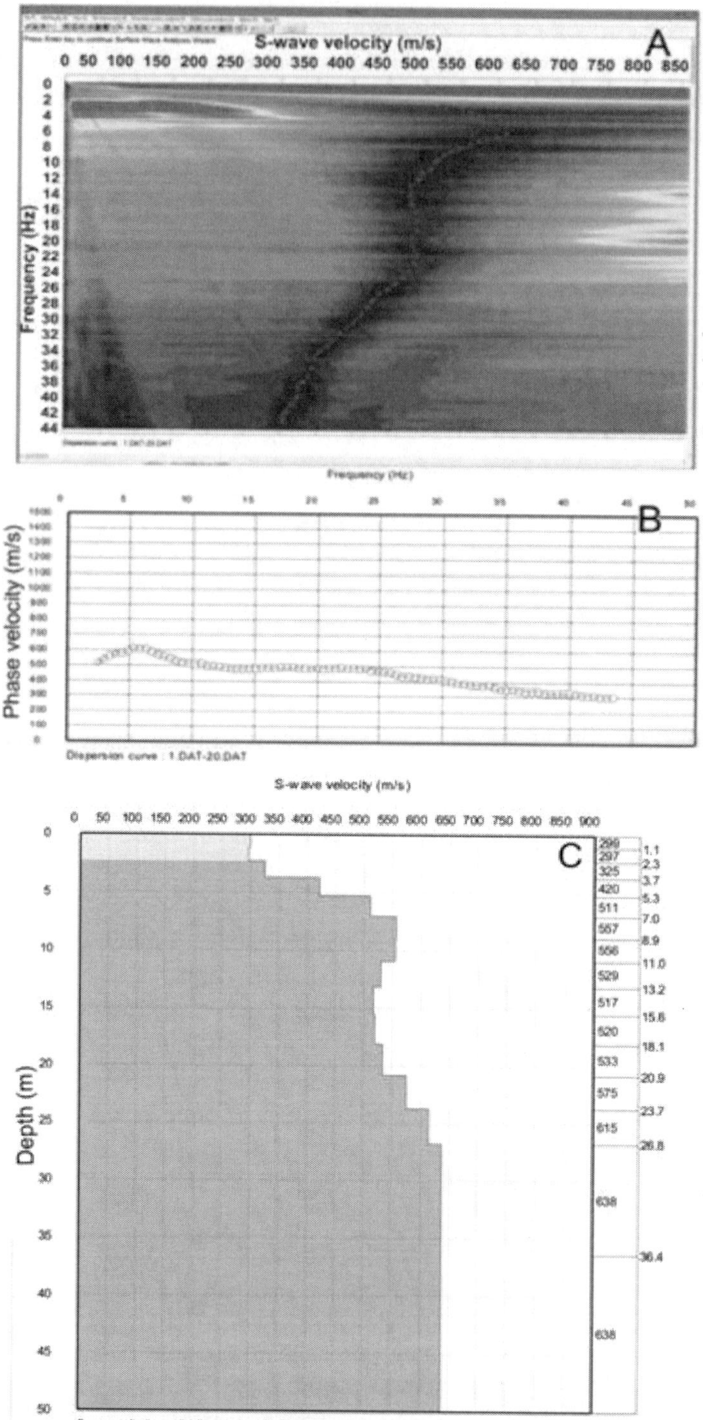

Figure 5. Processing steps of SASW data. A) Phase velocity-frequency plot (in color contours) is presented with the determined maximum amplitudes shown as red dashed line. B) The resulted dispersion curve is shown after processing of the phase velocity-frequency plot (A), C) Final Vs model after inversion of the dispersion curve as shown in (B).

determined for a quadripole of electrodes, the next set of four electrodes is automatically selected and a second measurement of resistance is made. This procedure is fully automated and the time for the data collection is minimized due to modern instrumentation. This process is repeated until the end of the line is reached. The line is then re-surveyed with an inter-electrode spacing of 2, 3, 4, etc. Each increase in inter-electrode spacing increases the effective depth of the survey. The measured resistance values are converted to the value of apparent resistivity, (in ohm-metres) which can then be used to model the true subsurface resistivity distribution. The same acquisition system can be used for collecting IP data. Specifically, the measurements are made of both the resistivity and chargeability of the subsurface. Measurements of self potentials are made with two non-polarisable porous-pot electrodes connected to a high impedance voltmeter. Traditional metal stakes, as used in resistivity surveying, cannot be used as they generate their own potential when they are inserted into the ground. Data are collected along a survey line (SP profiling) or across a grid to produce a contour map of self potentials. The data require processing, as most interpretations are based on qualitative analysis of profile shape, polarity and amplitude. In geoelectrical methods, the orientation of the profile is crucial to ensure a high quality of data avoiding the anisotropy effects of the background geological formations.

Processing and Interpretation of geoelectrical data. Initially, data points, corresponding to wrong resistivity values, are removed. Next, filtering of raw data includes the removal of noisy points. After these preliminary steps, the user can continue processing by changing the inversion and regularization parameters or methods. Detailed instructions and information on how to process geoelectrical data can be found in several sites and manual of commercial software.

In many cases, the resulted tomographic models can be unreliable due to following reasons:

- Every reliable model needs good data. If the electrodes are not plugged well in the ground due to stones, hard soil or other reasons, or if the upper part of the investigation site is very dry (resistive) and the contact resistance is high or very conductive, the expected quality of raw data will be low. Try to solve such problems by changing the acquisition method/parameters or the geophysical method (applying seismics) in the field before pressing the start button.
- Non-uniqueness of inverse models. Same response can be resulted from different subsurface models. Geophysical methods and inversion can provide mathematical solutions with low error between observed and calculated parameters. Sometimes, depending on the complexity of the study area, this resulted model has nothing to do with reality. We should always use other kind of information such as borehole logs or other methods to constrain the model and construct a reliable geological model.
- Complex 3D geology. For processing, the most of available software assumes a 2D subsurface model. This assumption is true when the measurements are taken in proper orientations. If the geology is very complex, a distortion of the deeper sections of the resulted model is expected. For that reason, 3D processing software was recently used to give us an idea of the extent of the three-dimensional effects.

3.3. Gravity Data Acquisition, Processing and Interpretation

Gravity measurements are recorded along profiles, crossing the possible location of the subsurface target. The data are acquired using a base station and the mobile instruments. Sometimes for profile/grid of small length/area, and if the second instrument is not available, only one instrument can be used for both base station measurements and profile/grid measurements. The gravimeter, located in the base station, has to be on stable ground, concrete or not and protected from the weather conditions. The data of the base station are used to correct the grid data for instrumental drift and earth tides. For wide surveys, the base station measurement (the mean value of some readings to within 10 μGal) should be taken at the start and end of the day. At each point, in the profile/grid coordinates must be recorded.

Gravity data have to be corrected for elevation, influence of tides, latitude and, in the case of local topography irregularities, for topography. Bouguer gravity anomaly is the difference between observed gravity and theoretical gravity on the Earth's surface, after making all the required corrections. This is the final product used for the interpretation [3].

According to [102] and [3] concerning the interpretation (Figures 2, 3) of a gravity anomaly:

- A positive gravity anomaly is due to feature of higher than average density.
- A negative gravity anomaly is due to feature of lower than average density.
- The dimensions of the target are proportional to the induced anomaly.
- A shallow feature induces a sharp high frequency anomaly.
- A deep feature induces a broad low frequency anomaly.
- The shape of the induced anomaly can be used to give more information on the depth, the shape and the total mass of the target.

3.4. Magnetic Data Acquisition, Processing and Interpretation

Magnetic measurements on ground are taken along a profile or in a grid at regular space interval, depending upon the target's depth and dimension. Shallow and small targets require surveys of higher resolution. The space interval is usually 0.5-1m and even smaller in very high resolution surveys. At each station, the mean value of generally 5-20 readings is taken. The magnetic response of a possible target is generally more complex compared with its gravity response due to the dipole nature of the magnetic field. This is clearly seen in Figures (2) and (6), prepared by the same software (pdyke), using the same field and model parameters, corresponding to a prism. However, in practice environmental magnetic data are often simpler to interpret than gravity because the noise and contributory sources are easier identified. The shape of a magnetic anomaly is often asymmetrical and varies depending on latitude and orientation of the causative body and the observed profile [3].

Magnetic susceptibility can be measured directly in the field or by collecting samples and placing them in plastic containers. Prior to starting the magnetic susceptibility measurements, the samples have to be mixed, air-dried, disaggregated and sieved, retaining the fraction smaller than 2mm in order to reduce the biasing effect of air, water and pebbles. Additionally, each sample is weighted and the susceptibility measurements in both frequencies are multiplied by a factor w_f = (10/weight of sample) in order to normalize the measurements for a mass of 10 gr.

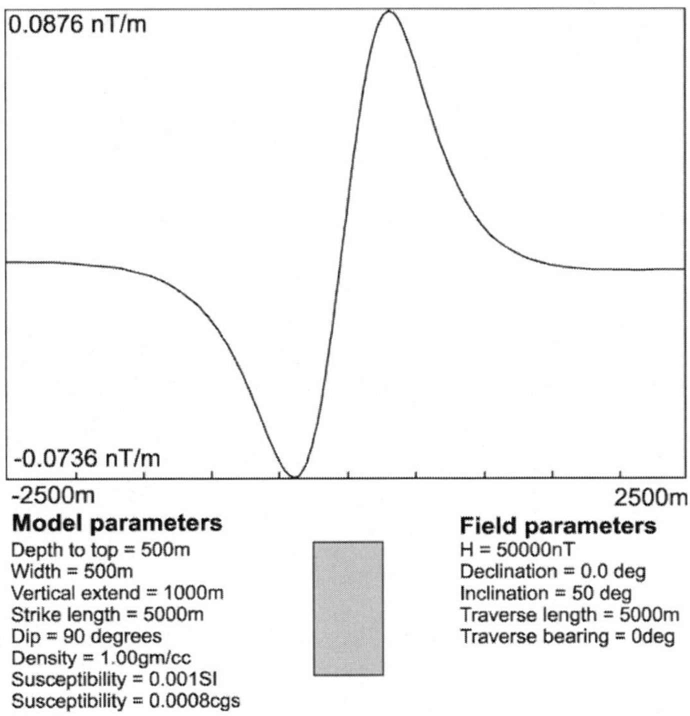

Figure 6. Vertical gradient of the magnetic field corresponding to a prism. Modeling has been done by the free software pdyke (Geophysical Software Solutions) [61].

Generally, the magnetic data are initially corrected for the survey parameters and noise. Diurnal correction is applied in the field measurements for the temporal changes in the Earth's field during the survey, using the repeating measurements in a base station at frequent intervals. In the case of magnetic storms, magnetic surveys should stop. Normal field correction includes variations in field with latitude and longitude, in case of very large surveys. Elevation and terrain corrections are applied in any kind of magnetic survey (small or large areas). Finally, the regional field and the inclination-declination effects (reduction to the pole) are also removed from the data.

Magnetic data resulting from mapping small areas, such as magnetic survey of archaeological sites, are initially processed to correspond to a common base level (0-level base line) for all grids. This is to eliminate the shift of average value in the grids, due to differences in balancing the instrument and shifting of base/reference stations. The change of coordinates and the correction factors are used to create a synthetic mosaic of the grids. Maps are created through interpolation techniques. Selective de-spiking techniques and compression of the dynamic range of values are employed to isolate anomalies close to the background level. Grid equalization and line equalization are very useful techniques to smooth the data and to avoid stripping effects. Other filters such as high-pass filters (gradient) or the calculation of first horizontal derivatives are helpful in emphasizing the high frequency components of the magnetic maps.

3.5. Electromagnetic Data Acquisition, Processing and Interpretation

Terrain Conductivity Meters (TCM) data acquisition processing and interpretation. TCM read the apparent conductivity of the study areas in profile mode or acquiring measurements in grid (every 0.5 or 1m in X, Y direction). The distance between measurements is called the station spacing. In general, the station spacing should be between one-third to one-half the intercoil spacing. As soon as the target is found, additional geophysical methods are applied to reconstruct and refine the final subsurface model. Measurements in different coil orientations and/or intercoil separations can provide valuable information about the conductivity change with depth.

TDEM data acquisition processing and interpretation. The penetration depth depends on the size of the antenna (the bigger the loop size, the deeper the investigation), and on the type of the rocks. Three TEM measuring array configurations are used: central loop, offset loop and the coincident loop (or single-loop). The coincident loop configuration is the more simple, and perhaps the most usable. The same wire loop is used for both transmitter (TR) and receiver (REC), and major advantages of this loop configuration include good signal-to-noise ratios and a uniform response as a function of transmitter location. However, a homogeneous relief area is required. During the TEM fieldwork, the operator should be very careful for the best possible site selection, the correct installation of the loop, and other conditions which can produce noisy and bad quality data, like man-made noise (power and telephone lines, pipelines, fences, etc).

After the raw data collection, the data should be edited, filtered and smoothed prior than the final modeling and imaging. Editing is a very important part of the analysis of the data, since the operator can correct mistakes contained in the header of the field data (Name, TR, REC, coordinates), can exclude wrong points from the resistivity curves and can change the position of a point. Once the resistivity data are derived, the next stage is modeling of the data, as the ultimate goal is to deduce the subsurface resistivity distribution (resistivity-depth information) from these sounding curves.

In Figure (7), an example of data smoothing is shown. Two more curves, except the resistivity curve, are given as result of the raw data smoothing. The first (upper) continuous curve approximates initial $\rho\alpha(t)$ data and the second (lower) curve is the calculated dependence $\rho full(t)$.

GPR data acquisition processing and interpretation. The basic signal processing of the GPR data includes: 1. Post processing gain, corresponding to amplitude corrections 2. DC bias removal, that is elimination of the data offset due to the electronic design, 3. De-banding through background removal, corresponding to removal of the signal mean value from every trace in the horizontal dimension, 4. Migration, in order to put the reflection points in their true location and 5. Hilbert transform corresponding to an estimation of the relative power within the signals.

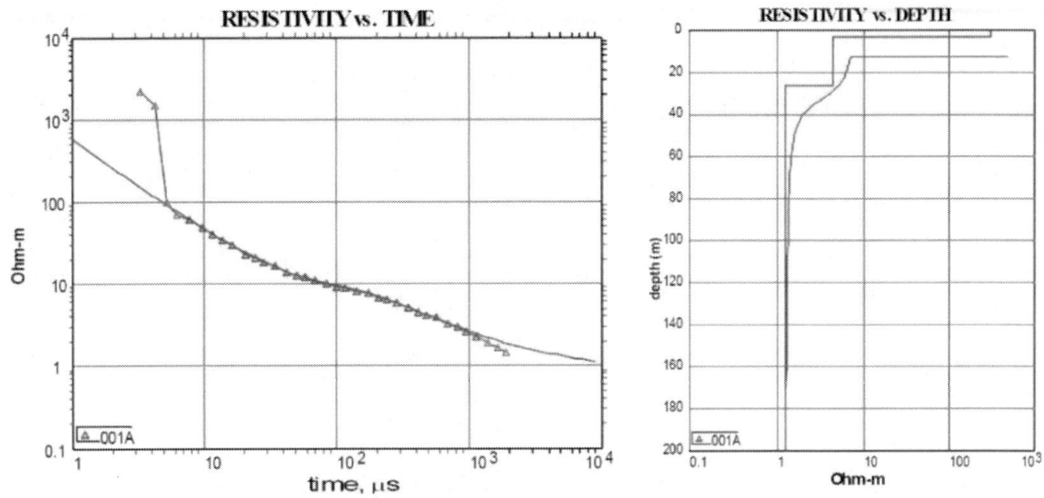

Figure 7. (Left) Curve of apparent resistivity ρa(t) versus time (example of sounding 001A). (Right) The transformation (ρ(h)) and inversion (example of sounding 001A) resistivity model with depth are presented through the smooth curve and the piecewise-uniform diagram, respectively (modified after [103]).

4. CASE STUDIES – DISCUSSION

The literature is abundant with applications and new techniques of data processing and interpretation of Environmental Geophysics, facing a variety of issues in the fields of the engineering, the pollution, the geological hazards and the archaeology. The main issue, arising prior to every survey, is the selection of the proper combination concerning the geophysical methods, which has already been analyzed in the previous paragraphs.

4.1. Geoenvironmental Monitoring by Using Geophysical Methods

Magnetic Measurements and Geochemical Analyses as a Pollution Monitoring Tool

Pollution monitoring [17], using a combination of the soil magnetic properties and geochemical analyses, was applied in the wide area of the power plant of Megalopolis (Peloponnesus, Greece). The magnetic behavior of the samples from the study area has been estimated by analyzing the spatial distribution of the magnetic susceptibility, the variation of the magnetic susceptibility with temperature, the anhysteretic remanent magnetization (ARM) and the isothermal remanent magnetization (IRM). The main factors of the pollution transmission were estimated to be the drainage network and the wind. Magnetic properties of the examined soil samples agreed that the metal pollution reveals high values around the quarries and depositories and is generally oriented along a NW-SE direction (Figure 8). Additionally, the pollution is transmitted northwestwards out of the Megalopoli's basin from the drainage network. High linear correlation factors were observed between Fe and Ni, K, Rb, Y concentrations respectively in sediments older than Pleistocene, indicating the relationship of the geologic features with the specific metals. Low correlation between Fe concentration and the rest of metals has been determined for the samples collected from sites filled by Holocene sediments. The terrain attributes played an important role in the distribution of the heavy metals, and in this term GIS techniques provided an excellent tool for studying the spatial distribution and relationship between magnetic susceptibility, heavy metal concentrations and the natural settings of a study area. This investigation gave rise to use magnetic susceptibility studies as a pollution monitoring tool around local power plants with a dense traffic net in Crete [68, 18].

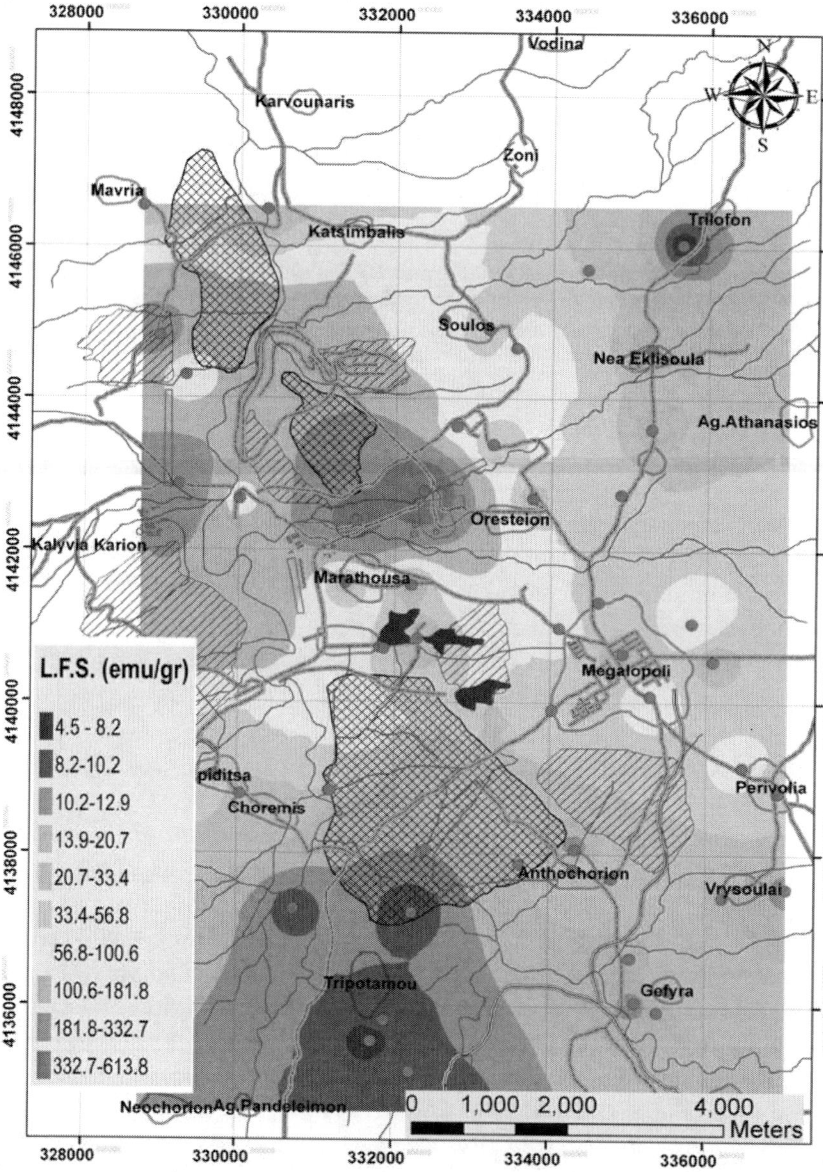

Figure 8. Mapping of the low field magnetic susceptibility (L.F.S.) in the wide area of the Megalopolis power plant (Peloponnesus, Greece; modified after [17]).

Geoelectrical Characterization of an Olive Oil Mill Waste (OOMW) Site

A lot is well known about the beneficial properties of olive oil but little is known that this virgin food produces huge amounts of waste. Olive oil mill waste (OOMW) has a dark brown color, unpleasant smell and high organic (phenols and polyphenols) and low inorganic compounds [104,105]. At present to our knowledge, there is no strict policy in many countries concerning OOMW disposal. The wastes are deposited freely in unprotected evaporation ponds. The landfills are usually close to rivers or over permeable geological formations. Thus, the vulnerability for groundwater contamination and soil degradation is high [106].

Electrical geophysical methods (ERI, TDIP, SP) can provide an efficient [105], non-invasive method to characterize and monitor environmental change/degradation.

Figure (9) on the top, shows the study area (OOMW). This site is used for studying the efficiency of electrical geophysical methods for monitoring the subsurface contamination. The area is mainly composed of alluvial deposits (with variable hydraulic permeability). The OOMW (red color ellipse in Figure 9, upper figure) with dimensions 75m by 25m is constructed in order to dispose the waste produced by the nearby olive oil factory.

The complete ERI survey involves data collection along 6 profiles but the results from only one profile (yellow arrow - Line 1 in Figure 9) are presented. Line 1 was chosen for continuous subsurface monitoring and three different geoelectrical measurements (ERI, TDIP and SP) were acquired. Different array configuration and electrode spacing were applied to optimize vertical and horizontal resolution and to estimate the error.

The acquisition of TDIP started on August 2014 and continued until December 2014, collecting in total 4 data sets (~ one data set per month). Five SP data sets were collected between May to July 2014 (every 2 weeks) using 24 Pb/PbCl non-polarized electrodes installed along Line 1. The collection of ERI data was started on October 2013 and continue till now acquiring a full dataset every month. We should mention that the ERI and TDIP data quality was excellent as evidenced by reciprocal measurements.

The resulted 2D geophysical ERI and TDIP profiles are presented in Figure 9. The high resistivity top layer, correlated with the thin consolidated/stiff surface layer composed of sands, pebbles and gravels. Towards the bottom of the profile a conductive anomaly is depicted (mainly the left side of the image) which is the OOMW plume. This assumption is verified by soil and liquid sampling from the study area and physical and chemical analyses of the collected samples. The OOMW produce a highly conductive plume that can be imaged applying all electrical methods (ERI and TDIP). The TDIP tomographic image shows that the plume is better depicted with imaginary conductivity data.

Moreover, along the line 1 a SP dataset was collected. A dataset of SP and ERI data recorded at the same date was selected and presented in Figure (10), showing a comparison between the two different methods. The SP profile is shorter that the ERI, as shown in Figure (10). The conductive (blue color) plumes at the first half of the ERI profile and is very clearly indicated and well determined also by the SP data, as shown in the Figure (10). The qualitative interpretation of the SP data, as suggested by the typical SP responses, presented in the left side of Figure (10), is in a good agreement with the resulted tomographic image of ERI. Thus, in the first half of Line 1, a downward flow is expected and confirmed by ERI results and the contaminant movement along unconfined aquifer is predicted at the second half of the profile.

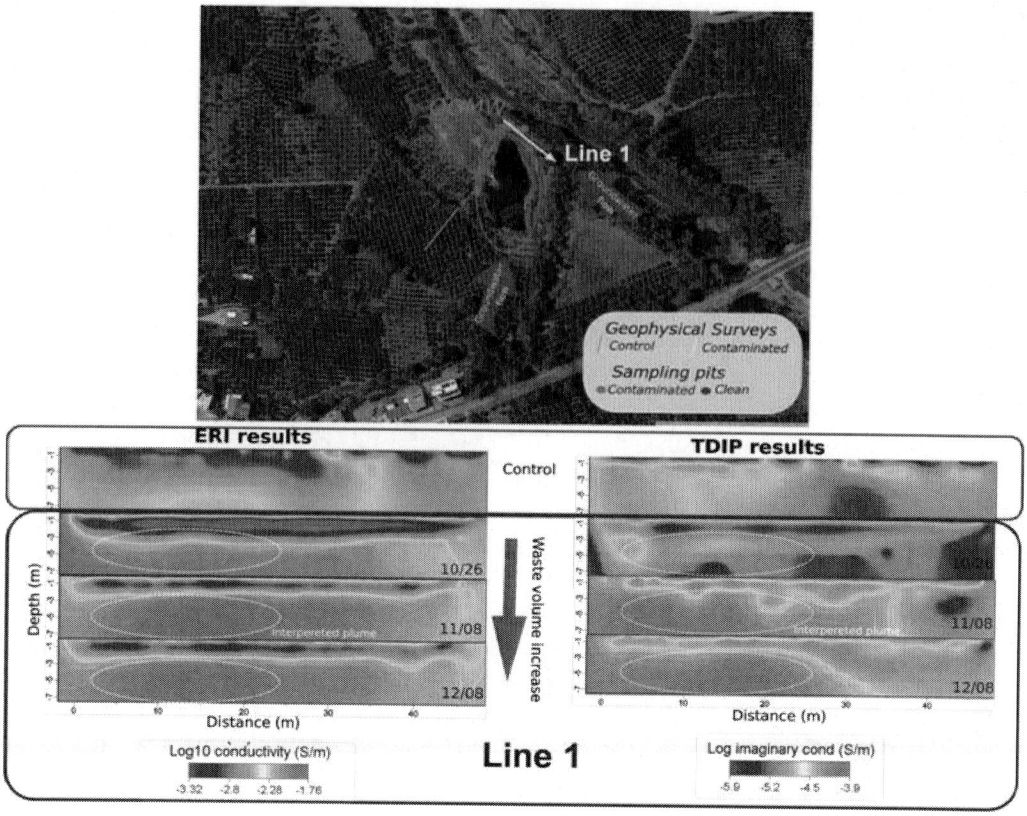

Figure 9. Location of the geoelectrical lines that cover the area of the OOMW site is presented on the top. Inverted images of the ERI (left) and TDIP (right) surveys. In the upper part, the responses from the uncontaminated site (control) are shown. In the lower section, the response of the contaminated area is shown at 3 different times, with increasing waste load (modified after [105]).

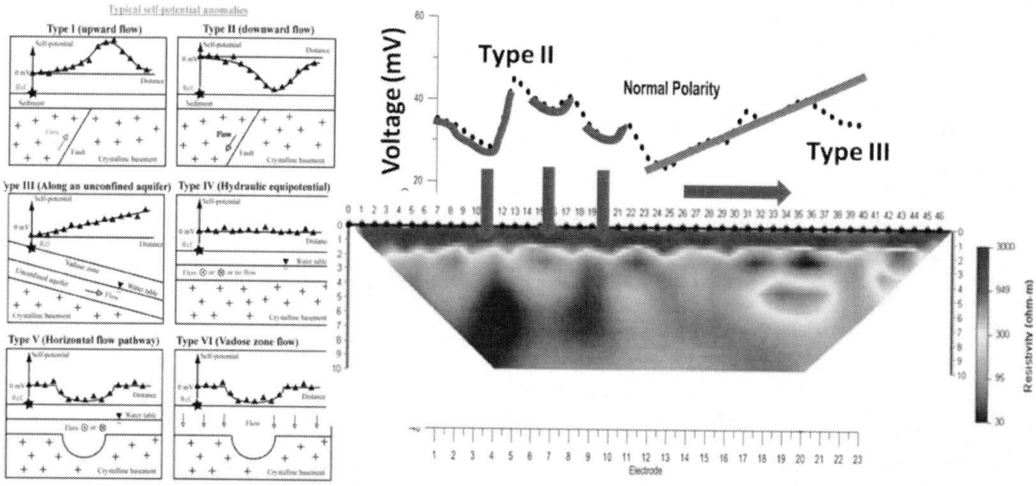

Figure 10. A comparison between the ERI and SP data set is shown in the right side of the figure. The expected response/anomalies of the SP signal, depending on the geological and hydrogeological conditions, are also presented on the left (modified after [107]).

Hydrogeophysical Characterization of Freatic or Karstic Aquifers

It's not always straightforward for field geologists to conclude about the hydrogeological status of a watershed. The difficulties increase when the study area is more complex, due to geological and tectonic settings. In such a case, special methods are required to solve particular hydrogeological problems [108]. Many geophysical methods have been used for groundwater resources investigation, but the electrical and electromagnetic methods are the most popular, since they have the greatest success and can be used for studying fresh and/or contaminated (saline water) aquifers around the world [109, 110, 111, 112, 113, 114, 115, 116, 117, 118, 119, 120, 121 and 78].

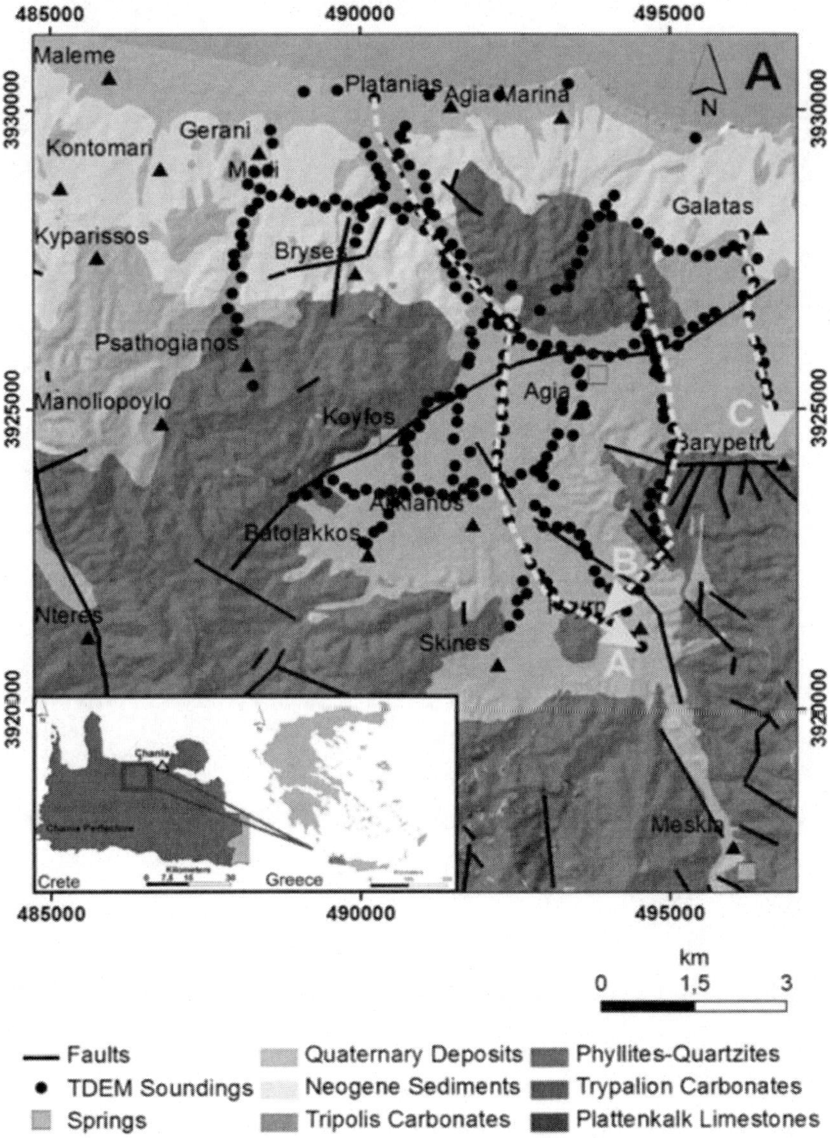

Figure 11. (Continued).

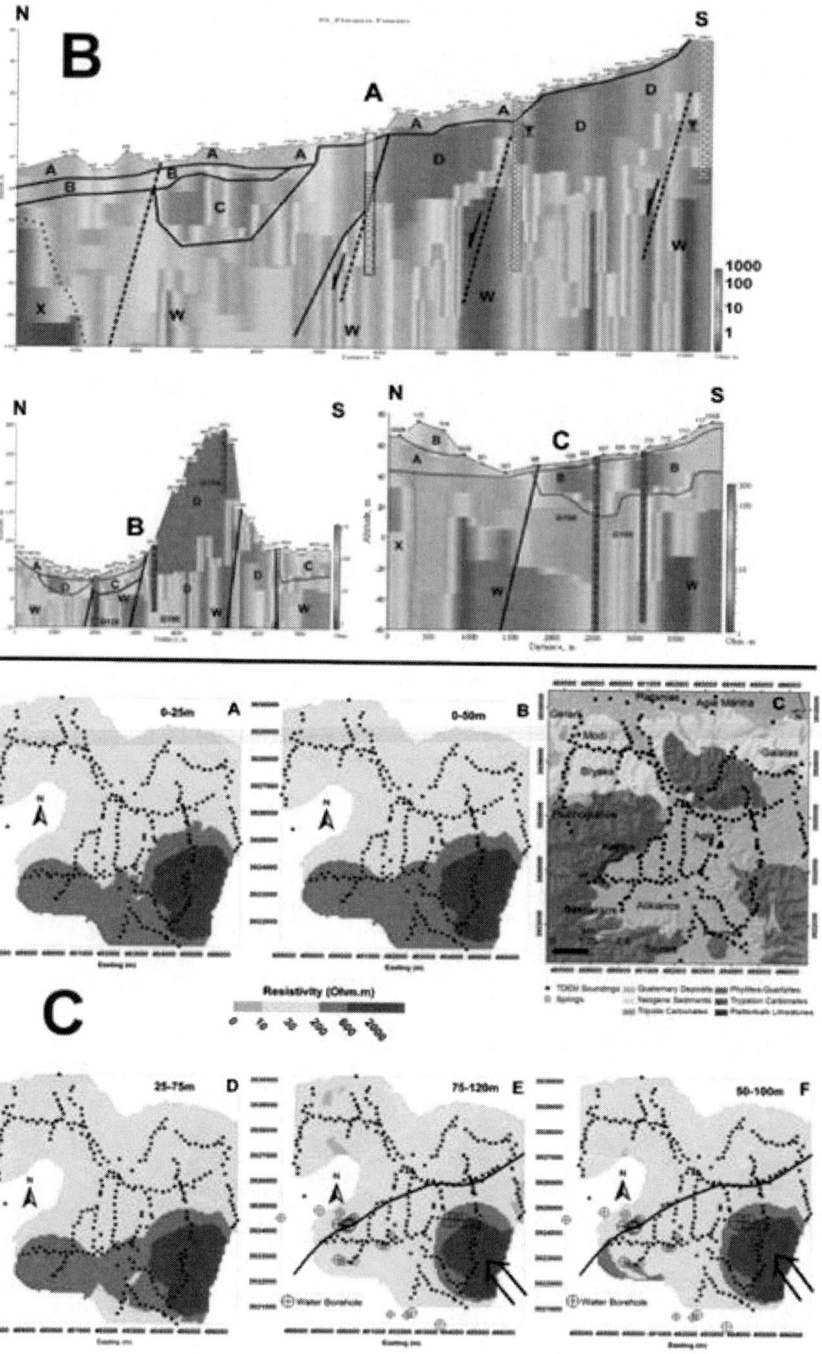

Figure 11. A) The location of the study area in different scales is presented. A high quality geological and tectonic map of the study area is also provided. Different colors demonstrate the several geological formations. The locations of the TDEM soundings are also depicted as black solid circles. Yellow dashed arrows present the 2D geophysical sections (A, B and C) from north to south. B) Three geoelectrical sections (A, B, C) are presented. C) The pseudo-3D geoelectrical interpretation till the depth of 120 m (five depth slices, A, B, D, E and F) is presented (modified after [103]).

In Figure (11), a large scale hydrogeophysical survey is presented, in order to define the current hydrogeological status of the Keritis Basin of NW Crete Island in Greece, by means of electromagnetic geophysical method. Specifically, transient electromagnetic (TDEM) measurements (soundings) were conducted in the Keritis basin, to obtain detailed information about the tectonic and hydrogeological characteristics of the aquifer of the area under investigation. The study area is composed of Quaternary deposits, Neogene sediments, Tripolis carbonates, Phyllites–Quartzites, Trypalion carbonates and Plattenkalk limestone (Figure 11). Solid black lines (Figure 11A) denote visible/concealed faults as depicted from TDEM modeling. A total of 919 geophysical measurements (sounding) were acquired in 314 different locations, in correlation with other related available information (such as geology, borehole logs, and tectonic) and finally, the 2D (Figure 11B) and pseudo-3D (Figure 11C) model reconstruction of the subsurface of the study area was achieved. These pseudo-2D sections were in agreement with the geological sections constructed from the geological map of the study area. The available boreholes logs were used for evaluation and confirmation of the resulted tomographic images. The tectonic features (fracture zones) as depicted from the processing, imaging and interpretation of the pseudo-2D sections (Figure 11B) are presented as thick solid and dashed black lines. The black thick line (as shown in Figure 11C at the deeper slices) depicts the main fracture zone in the study area (from WSW-ENE). Black thick arrows (as shown in E and F depth slices) define the direction of the groundwater flow through the study area. The groundwater flow model and the final reconstructed tomographic hydrogeophysical models can be used by the water management authorities and any future groundwater investigation and water management plan.

4.2. Use of Geophysical Methods to Solve Engineering Problems

Geophysical methods can be used in a wide range of geotechnical problems ranging from building ground investigations to the inspection of dams, roads and dikes [122, 123, 124, 125, 126, 127, 128, 129, 130]. The main purpose of every geo-engineering survey is the exploration of the subsurface geotechnical condition (geology, tectonic or indirect determination of physical parameters of the soil/rock) prior any construction.

Application of the Environmental Geophysics in Road Construction Works

The combination of detailed geological and tectonic mapping, seismic refraction, frequency domain electromagnetic and ground penetrating radar was applied to detect subsurface karstic voids and heterogeneities, resulting in dangerous collapse of road segments overlying these features [10]. The technique of 2D resistivity imaging was excluded, because subsurface fuel pipes were present.

Triassic – Cretaceous limestones, belonging to Tripolis Geotectonic Zone, crop out in the study area. These limestones are strongly karstified and most of the karstic voids are filled with Quaternary "Terra rossa." Geological mapping indicated two tectonic zones in the study area, mainly NW–SE striking, showing dips in the range of 50°-90°. The fault zones were mainly estimated by the seismic refraction, while EM conductivity and GPR determined finer features including faults and karstic cavities, generally greater than 50cm in diameter. The georeferenced geophysical maps were cross-correlated and areas of highly disturbed stratigraphy were marked (Figure 12). The tectonic action and the anthropogenic activity are mostly responsible for the highly disturbed stratigraphy. The eastern part of the road (Figure 12) is mostly affected by tectonic activity (karstic voids and tectonic zones), situated at depths ranging between 0.7 – 2.8m.

The proposed combination has provided a successful tool for the detection of highly disturbed stratigraphy. The detailed geological and tectonic mapping, prior the geophysical prospecting mapping, was a key asset, which dramatically reduced the time and the cost of the geophysical survey.

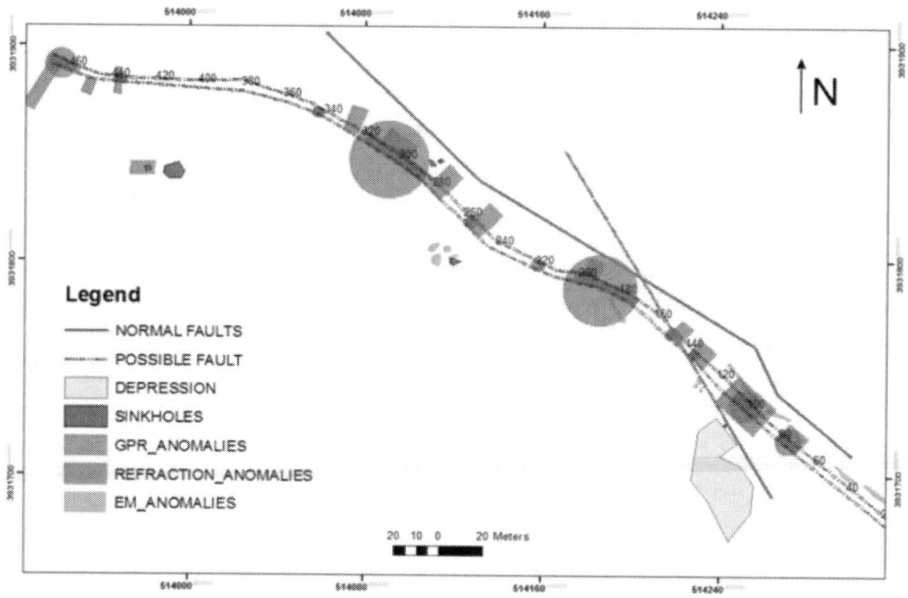

Figure 12. Map showing the combined interpretation of the applied methodology (Refraction, GPR, EM) for the detection of subsurface karstic voids and faults in road construction works (modified after [10]). Disturbed stratigraphy is mainly indicated in locations of the overlapped Refraction, GPR and EM anomalies.

Geophysics and Building Foundations

Several direct and indirect methods for obtaining subsurface information prior to any construction have been applied by [131]. Specifically, at the beginning of the project, the research team performed topographic map interpretation and the existing geological reports, maps and the available borehole logs from the broader area were studied. After that, geological mapping with in situ examination of the geological formations and their geotechnical characteristics (hard/stiff or not) has been taken place. Knowing the geological setting and having some knowledge about the geotechnical characteristics of the study area, two different geophysical methods were applied. Five parallel black solid arrows (Lines 1-5 in Figure 13A) together with the two perpendicular ones (Lines 6 and 7) indicate the electrical resistivity imaging (ERI) which were carried out in a 8m X 20m grid while the electromagnetic terrain conductivity (TCM) was collected in a grid of 20m× 20 m acquiring 1681 measurements (every 0.5m in X and Y) (Figure 13A). The ERI and TCM final resulted models were combined and presented into the colored dashed rectangular almost in the middle of the study area as is shown in Figure (13A, B).

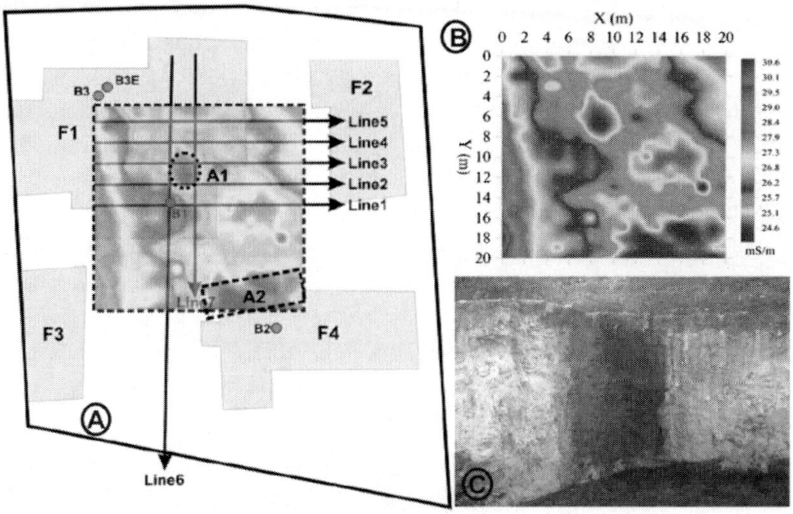

Figure 13. A) A topographic scheme of the study area is presented. Red filled circles (B1, B2, B3, B3E) depict the geotechnical boreholes and the dashed rectangle in the centre of the study represents the area which was scanned using the TCM method (as shown in B). Grey shaded polygons (F1–F4) show the places where buildings are going to be constructed. B) 2D contour image of the TCM measurements. Measurements of conductivity in mS/m. C) Excavation of the study area and the confirmation of the geophysical model is presented (modified after [131]).

A low geoelectrical resistivity anomaly from the ERI survey is depicted as a dashed ellipse (A1) and a low (ERI) resistivity anomaly which can be correlated with a destroyed stone wall found in the study area, presented as a dashed rectangle (A2) in Figure (13A). Resistivity data were processed by using the Res2DInv and Res3DInv commercial software [132, 133, 134]. The geoelectrical tomographic models were validated through the corresponding electromagnetic results. The final resulted electrical and electromagnetic model, was combined with the available geological and geotechnical information of the area under investigation. After all, the 3D subsurface model of the study area was reconstructed. The final integrated geophysical model was confirmed during new geotechnical boreholes and excavations were constructed. Later excavation works revealed a disrupted/discontinuous part of the marly limestone, filled with clay (conductive materials). The location of the geological–geotechnical inhomogeneity is in agreement with the resulted anomaly depicted by both geophysical engineering methods. The added value of the geophysical techniques was that the responsible engineer selected the appropriate footing of the structure based on the final geophysical model.

Use of Geophysics for Wind Turbines Foundation

In 2013, seven seismic profiles were acquired using the MASW method to characterize a site where seven wind turbines would need to have been installed in 2014. Based on the literature, the MASW is the most powerful non-destructive fast geophysical method for collecting seismic data. The Geometrics R24 Strataview Seismographs used for this survey using 24 low frequency (4.5 Hz) geophones. The record length was 2048ms. A 5Kg sledgehammer with a trigger mount on the handle used for producing the pulses/signal on the ground (Figures 14 and 15). The geophones are set up in a line at fixed locations and the shot is moved through the spread. The first shot is located off-end at a near offset of one-half the geophone interval. The shot is then advanced at an increment equal to the geophone interval so subsequent shots are located midway between geophones. As the shot number increases, the shot location advances by one interval across the survey distance. The last shot is located off the opposite end by the same near offset of one-half the geophone interval. The data were processed with the SeisImager/SW commercial software.

The data were collected following the same acquisition parameters and survey geometry but with different directions, since the direction of the survey line depends on the availability of the area of measurement after the excavation of the foundation site. The data were processed following the optimum processing steps. After all, the final 2D S-velocity sections were produced for all sites. The final velocity models were estimated after 6 to 10 iterations and the RMS in each site was less than 6% (or 67 m/s). Overall, the resulted velocity models were reliable and of high-quality.

Figure 14. The acquisition system, Geometrics Strataview R24 seismograph is shown.

Figure 15. The geophones used as well as the survey line for site 2 is presented.

Interpretation

For the interpretation of the resulted Vs models, a geological section passing through the sites where the wind turbines are going to be founded, is constructed (Figure 16). Next step was to import all the sites on the geological sections in order to correlate the resulted Vs model with the geological formations of the investigated sites. It should be mentioned that the geological section based on the 1:50.000 geological map of the IGME (Institute of Geology and Mineral Exploration of Greece). In this case, a detail geological mapping should be done for being able to have "safe" results. What we expect is to possibly find the contact (thrust zone) between different formations. The ophiolites are older and overlie the flysch formation.

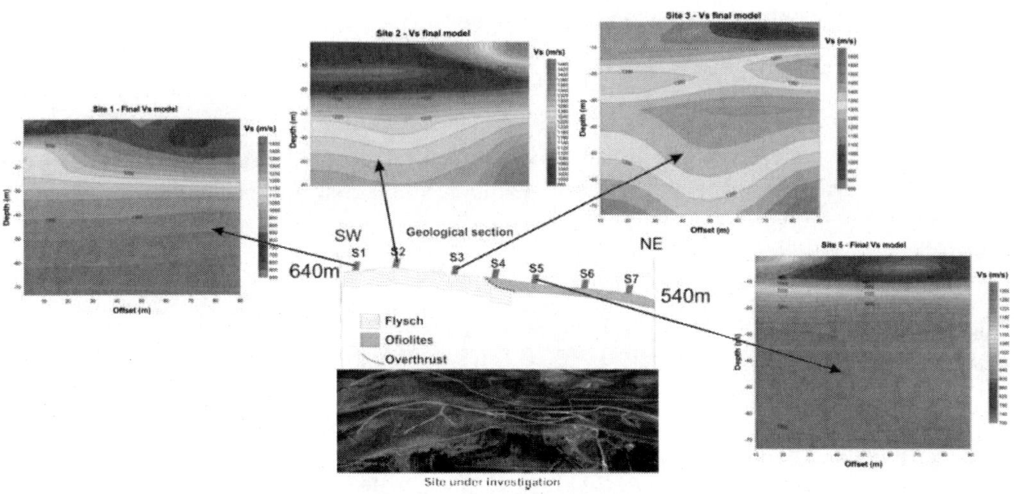

Figure 16. An overall interpretation of the MASW seismic survey incorporating the available geological information is presented. A geological section of the area under investigation is also presented on the bottom of the figure. Two geological formations were found. Ophiolites (grey shaded) and flysch. A thrust front located between the two geological formations is also shown.

As shown in Figure (16), the most interesting site is Site 3, this is due to a low velocity layer depicted at the average depth of 35 meters. The only geological explanation is that the low velocity layer is the flysch formation, which is located beneath the ophiolitic formation. Based on the results of the geophysical research, the thrust zone is possibly located west of that shown in the geological maps of IGME. The aforementioned information was used by the civil engineers for designing the foundation and construction of the wind-energy park in study area.

4.3. Archaeological Prospection Using Modern Geophysical Techniques

In the Kenchreai Cemetery Project (KCP) electromagnetic, magnetic, ground-penetrating radar and gravity techniques were applied to explore a major cemetery of Roman date in southern Greece [35]. Geophysical prospecting revealed, not only considerable structural and burial density in this transitional area on the periphery of the urban settlement, but also variation in the date and purpose of occupation on the ridge. Such studies contribute more generally to the discipline of funerary archaeology through experimentation on the efficacy of the survey methods and an innovative contextual approach to exploring the mortuary landscape.

Additionally, different geophysical techniques were applied in the course of a multidisciplinary research conducted within the framework of the Sikyon survey project [36], whose goal is the study of the landscape and human activity on the plateau of ancient Sikyon (NE Peloponnesus). Among the most impressive findings was a Basilica with an inner and outer narthex, perhaps from the early Byzantine, detected 1.5 m below the surface by the magnetic (Figure 17a), ERT and GPR (EKKO 1000) techniques.

Data interpretation represents the final step of the geophysical prospecting, because all collected information is combined to characterize the near-surface structure and to identify targets. Automatic enhancement and identification of partial curvilinear structures in geophysical images (Figure 17b) has been proposed [99]. A rotation and scale invariant filter and a pixel labeling method have been applied, providing a robust enhancement and detection of mostly line structures in 2D grayscale geophysical images from archaeological sites in Greece. The linear patterns of the subsurface architectural structures have been detected with considerable accuracy in space and with low computational cost. Experimental results on real and synthetic images and comparison with existing methods in the literature indicated the reliable performance of the proposed scheme.

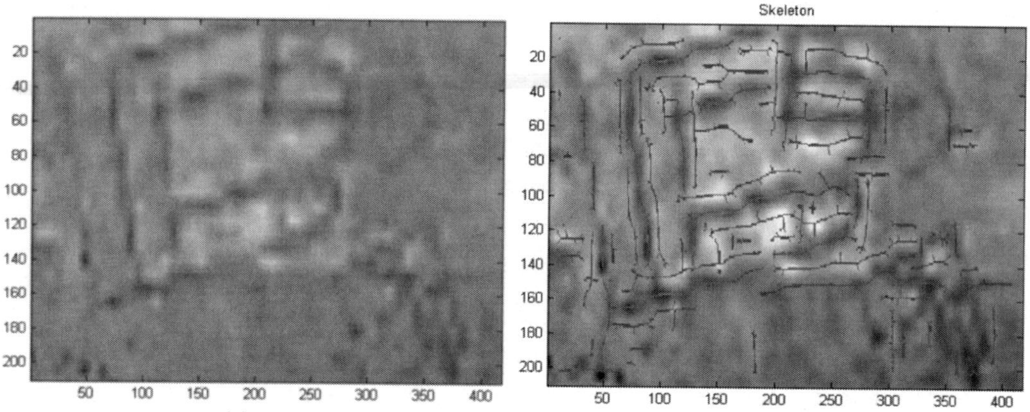

(a) (b)

Figure 17. (a) Vertical magnetic gradient measurements showing a Byzantine Basilica in ancient Sikyon, (b) Automatic detection of the linear patterns of the Basilica [36,99].

Shahrukh et al. [135] applied seismic refraction and geoelectrical tomography to determine the subsurface characteristics of the Priniatikos Pyrgos archaeological site in Eastern Crete. Main aim of this study was to locate the old natural harbour of the archaeological site. To do that, seismic and geoelectrical tomographic measurements have been combined to reconstruct the morphological characteristics of the basement in the study area and indirectly conclude the location of the buried harbour [136, 137, 138, 34, 139].

Twenty-nine refraction profiles, six electrical resistivity tomography sections and several geological cores (C#) and trenches (T#) were acquired and constructed to determine the soil thickness and bedrock geometrical characteristics (Figure 18A). First arrivals were picked and a 3D algorithm of seismic tomography was used to invert the travel times (Figures 18 B, C). Dashed lines in Figure (18B) depict a possible tectonic graben since a discontinuity of the geological layers was found. The ERI data were used to confirm the resulted seismic tomographic images (Figure 18D). The dashed black crooked line (Figure 18D) depicts the layers' interfaces for all seismic profiles and superimposed the resistivity models for better comparison. Both methods for the same area show comparable results.

The joint application and interpretation of seismic and resistivity tomographic data contributed to the establishment of specific models regarding the theories about the location of the natural canal/harbour [135].

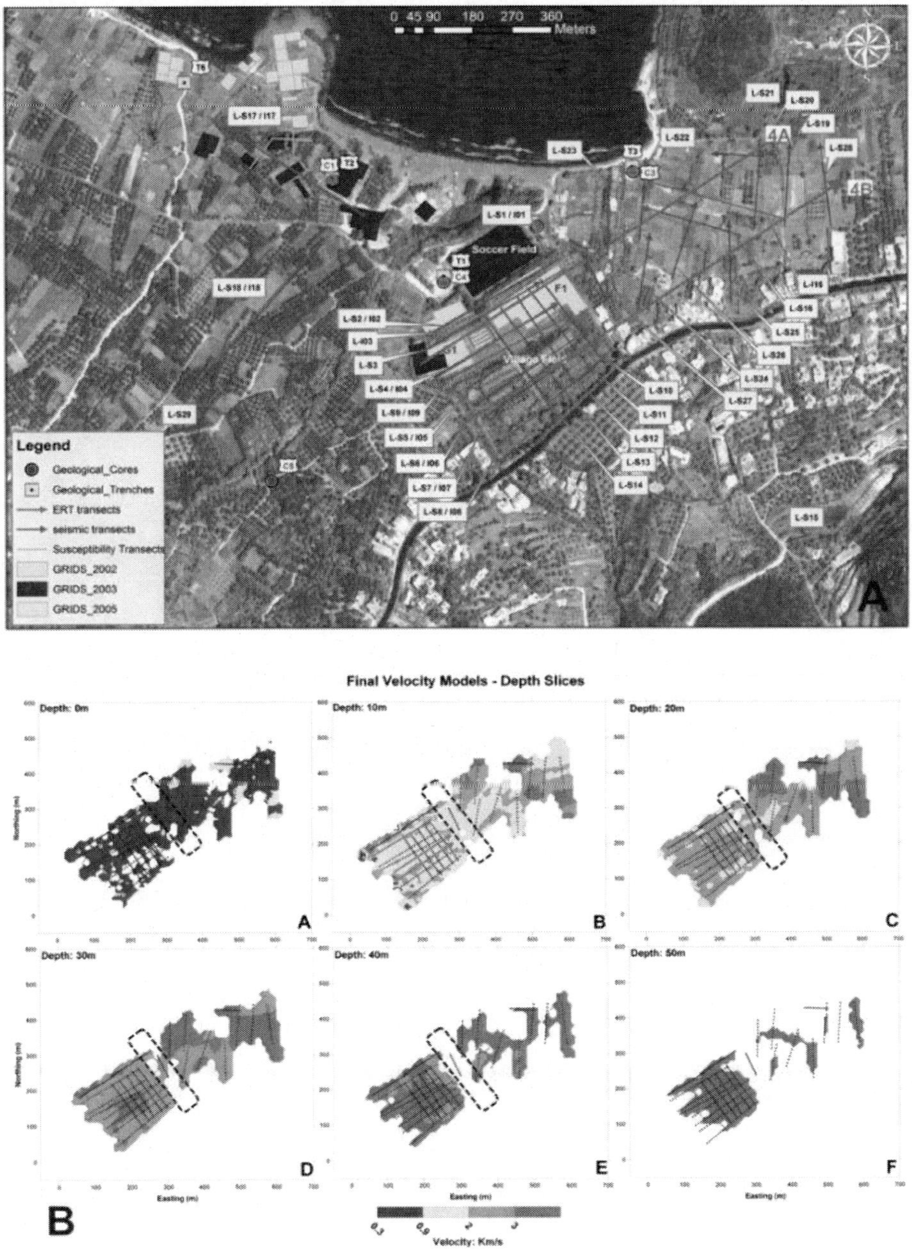

Figure 18. (Continued).

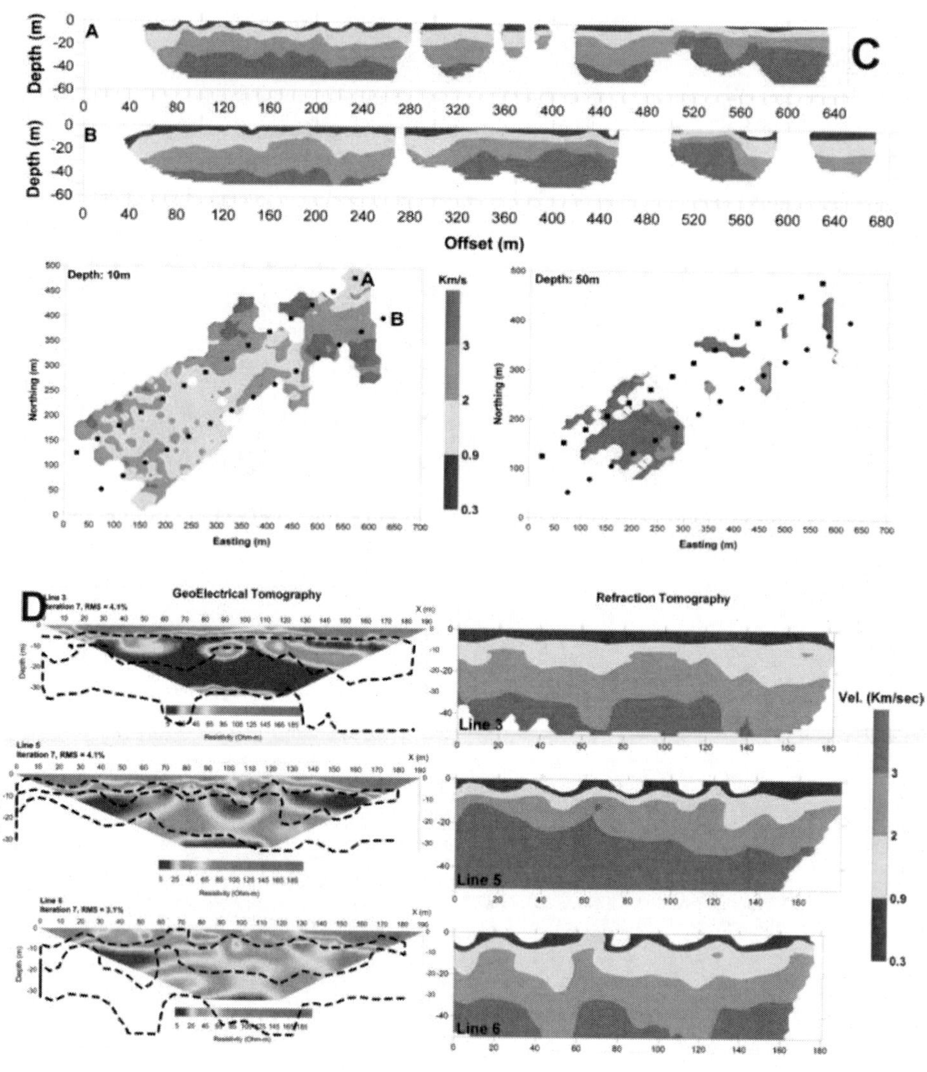

Figure 18. A) Locations of the ERI and refraction seismic profiles are presented. The location of the vertical tomographic images (C-A, C-B) are also depicted as solid green arrows. B) Final tomographic velocity models at different depths are presented. Dotted lines indicate the location and the refraction profiles. C) Cross-sections (A and B) of velocity distribution in depth in the study area are presented (shown also as green solid lines in A). Moreover, depth tomographic velocity slices at the depths of 10 and 50 m are shown. D) The comparison between the seismic and resistivity models with depth for Lines 3, 5 and 6 is presented (modified after [135]).

CONCLUSION

In the context of the present work the most familiar techniques in Environmental Geophysics are presented. This branch of Geophysics is intensively involved in the last 30 years in civil engineering industry, pollution studies and in archaeology. For each method, the basic principles, the survey planning, the data acquisition, the processing, the interpretation and the limitations of the pre-mentioned methods have been analyzed. Representative case studies, concealing the application of the Environmental Geophysics are also presented in order to show how the subsurface is monitored and features of interest are estimated with a considerable accuracy.

In concluding:

- Environmental Geophysics contributes to high-quality subsurface site characterization, using robust and effective tools to face a wide range of technical problems: 1. in civil engineering to explore the subsurface heterogeneities, fractures and voids resulting in dangerous collapses, as well possible problems arising from human activities, 2. in detecting toxic or potentially toxic substances for the public health, 3. in mapping archaeological sites to accurately determine the locations of the archaeological relics in order to preserve the human heritage.
- The techniques of the Environmental Geophysics are non-destructive, environmental friendly and cost effective.
- Each case in Environmental Geophysics is unique, because every region presents geological and technical particularities. So in order to be the survey effective, each time the appropriate combination of techniques must be selected. A well-planned environmental geophysical investigation is likely to achieve its scope. It is extremely important to determine the possible limitations and problems arising in every survey and always to take into account information associated to geology, hydrogeology and infrastructure.

Nevertheless, the results of an Environmental Geophysical investigation are not always as expected, and experts of other disciplines often become skeptic. So, it is always better to explain from the very beginning the limitations and the possible failures of a geophysical survey.

Acknowledgments

The authors wish to thank Dr. Francesco Sdao, Dr. George Vargemezis and Dr. Antonis Giannopoulos for their suggestions for the improvement of this paper. This project is implemented through the Operational Program "Education and Lifelong Learning," Action Archimedes III and is co-financed by the European Union (European Social Fund) and Greek national funds (National Strategic Reference Framework 2007 - 2013), Project Title: Interdisciplinary study for exploring, understanding and management of groundwater resources. Pilot field investigation North-Western and Central Crete (AQUADAM).

References

[1] Sharma, P. V. Environmental and Engineering Geophysics. Cambridge University Press, 1997.

[2] Annan, A. P. (1997) Engineering and environmental geophysics: the future. Geological Society, London, *Engineering Geology Special Publications,* 12(1), 419–426.

[3] Styles, P. Environmental Geophysics. EAGE Publications, 2012.
[4] Baranov, W. (1975) Potential fields and their transformations in applied geophysics; *Geoexploration Monographs*, 1(6), 121p.
[5] Owen, T. E. (1983) Detection and mapping of tunnels and caves. In: A. A. Fich, Ed.; Developments in Geophysical Exploration Methods; Applied Science Publishers: 5. Barking, Essex.
[6] Cripps, J. C.; McCann, D. M.; Culshaw, M. G.; & Bell, F. G. (1988) The use of geophysical methods as an aid to the detection of abandoned shallow mine workings. In: MINESCAPE; Harrogate; Institute of Mining Engineers, 281-289.
[7] Daniels, J. (1988) Locating caves, tunnels and mines. *The Leading Edge,* 7, 35–52.
[8] Hinze, W. J. (1990) The role of gravity and magnetic methods in engineering and environmental studies. In: Geotechnical and environmental geophysics; Ward, S. H., Ed.; *Society Exploration Geophysicists*, 1, 75-126.
[9] Wightman, W. E.; Jalinoos, F.; Sirles, P.; & Hanna, K. (2003) Application of Geophysical Methods to Highway Related Problems; Federal Highway Administration, Central Federal Lands Highway Division, Lakewood, CO; FHWA-IF-04-021.
[10] Kokinou, E.; & Sarris, A. (2011) Detection of the near surface structure through a multidisciplinary approach. *Central European Journal of Geosciences*, 3(4), 349-357.
[11] Benson, R.; Glaccum, R.; & Beam, P. (1981) Minimizing Cost and Risk in Hazardous Waste Site Investigations Using Geophysics. National Conference on Management of Uncontrolled Hazardous Waste Sites: Washington D. C.
[12] Frischknecht, F. C. (1900) Application of Geophysical Methods to the Study of Pollution Associated with Abandoned and Injection Wells. In: Doe, B. R., Eds.; Proc. of a U.S. Geological Survey Workshop on Environmental Geochemistry, U.S. *Geological Survey Circular*, 1033, pp 73-77.
[13] Wilson, S. R.; Ingham, M.; & McConchie, J. A. (2006) The applicability of earth resistivity methods for saline interface definition. *Journal of Hydrology*, 316, 1–4, 301–312.
[14] Georgaki, I.; Soupios, P.; Sakkas, N.; Ververidis, P.; Trandas, E.; Vallianatos, F.; & Manios, T. (2007) Evaluating the Use of Electrical Resistivity Imaging Technique For Improving CH4 and CO2 Emission Rate Estimations in Landfills. *Science of the Total Environment*, 389, 522-531.
[15] Soupios, P.; Papadopoulos, I.; Kouli, M.; Georgaki, I.; Vallianatos, F.; & Kokkinou, E. (2007a) Investigation of Waste Disposal Areas Using Electric Methods: A Case Study from Hania – Crete, Greece. *Environmental Geology*, 51(7), 1249-1261.
[16] Soupios, P.; Papadopoulos, N.; Papadopoulos, I.; Kouli, M.; Vallianatos, F.; Sarris, A.; Manios, Thr. (2007b) Application of Integrated Methods in Mapping Waste Disposal Areas. *Environ Geol*, 53(3), 661-675.
[17] Sarris, A.; Kokinou, E.; Aidona, E.; Kallithrakas-Kontos, N.; Koulouridakis, P.; Kakoulaki, G.; Droulia, K; & Damianovits, O. (2009) Environmental study for pollution in the area of Megalopoli power plant (Peloponnesus, Greece). *Environmental Geology*, 58, 8, 1769-1783.
[18] Kokinou, E.; Belonaki, C.; Sakadakis, D.; & Sakadaki, K. (2013) Environmental monitoring of soil pollution in urban areas (a case study from Heraklion city, central

Crete, Greece); In: Bulletin of the Geological Society of Greece, Proceedings of the 13th International Congress, 2, pp 963-971.

[19] Seferou, E.; Soupios, P.; Kourgialas, N.; Dokou, Z.; Candasayar, E.; Papadopoulos, N.; Karatzas, G.; Sarris, A.; & Sauter, M. (2013) Olive oil mill wastewater transport under unsaturated and saturated laboratory conditions using geoelectrical resistivity tomography method and the FEFLOW model. *Hydrogeology Journal*, 21(6), 1219-1234.

[20] Ntarlagiannis, D.; Robinson, J.; Soupios, P.; & Slater, L. Field-scale electrical geophysics over an olive oil processing site: evaluating the information content of resistivity versus induced polarization (IP) images for delineating the spatial extent of organic contamination, *Journal of Applied Geophysics*, in review, 2015.

[21] Power, C.; Gerhard, J. I.; Tsourlos, P.; Soupios, P.; Simyrdanis, K.; & Karaoulis, M. (2015) Improved time-lapse electrical resistivity tomography monitoring of dense non-aqueous phase liquids with surface-to-horizontal borehole arrays. *Journal of Applied Geophysics*, 112, 01, 1-13.

[22] Seaton, W. J., & Burbey, T. J. (2002) Evaluation of two-dimensional resistivity methods in a fractured crystalline-rock terrane, *Journal of Applied Geophysics*, 51, 1, 21–41.

[23] Wood, R.; & Curtis, A. (2004) Geological prior information, and its applications to solving geoscientific problems. Geological Society, London, *Special Publications*, 239(1), 1–14.

[24] Styles, P.; Toon, S.; Thomas, E.; & Skittrall, M. (2006) Microgravity as a tool for the detection, characterization and prediction of geohazard posed by abandoned mining cavities. *First Break*, 24(5), 51–60.

[25] Kouli, M.; Loupasakis, C.; Soupios, P.; & Vallianatos, F. (2009) Landslide Hazard Zonation in High Risk Areas of Rethimno Prefecture, Crete Island, Greece, *Natural Hazards*, 52 (3) 599-621.

[26] Lazzari, M.; Loperte, A.; & Perrone (2010) A. Near surface geophysics techniques and geomorphological approach to reconstruct the hazard cave map in historical and urban areas. *Advances in Geosciences*, 24, 35–44.

[27] Zhu, J.; Currens, J. C.; & Dinger, J. S. (2011) Challenges of using electrical resistivity method to locate karst conduits—a field case in the Inner Bluegrass Region, Kentucky. *Journal of Applied Geophysics*, 75, 3, 523–530.

[28] Saltus, R. W.; & Blakely, R. J. (2011) Unique geologic insights from "non–unique" gravity and magnetic interpretation. *GSA Today*, 21(12), 4–11.

[29] Moisidi, M.; Vallianatos, F.; Soupios, P.; & Kershaw, S. (2012) Spatial spectral variations of microtremors and electrical resistivity tomography surveys for fault determination of southwestern Crete (Greece), *Journal of Geophysics and Engineering*, 9(3), 261-270.

[30] Moisidi, M.; Vallianatos, F.; Soupios, P.; Kershaw, S.; Rust, D.; & Piscitelli, S. (2013) Modeling tectonic features of the Kissamos and Paleohora areas, Western Crete (Greece): combining geological and geophysical surveys. *J. Geophys. Eng.*, 10(2), 025015.

[31] Vafidis, A.; Soupios, P.; Economou, N.; Hamdan, H.; Andronikidis, N.; Kritikakis, G.; Panagopoulos, G.; Manoutsoglou, E.; Steiakakis, M.; Candansayar, E.; & Schafmeister,

M. (2014) Seawater intrusion imaging at Tybaki, Crete, using geophysical data and joint inversion of electrical and seismic data. *First Break*, 32, 107-114.

[32] Goodman, D. (1994) Ground–Penetrating Radar Simulation In Engineering And Archaeology. *Geophysics,* 59 (2), 224–232.

[33] Papadopoulos, N. G.; Sarris, A.; Kokkinou, E.; Wells, B.; Penttinen, A.; Savini, E.; Tsokas, G. N.; & Tsourlos, P. (2005) Contribution of multiplexed electrical resistance and magnetic techniques to the archaeological investigations at Poros, Greece. *Archaeological Prospection,* 13, 2, 75-90.

[34] Sarris, A.; Kouriati, K.; Kokkinou, E.; Aedona, E.; Karagianni, L.; Vargemezis, G.; Stamatis, G.; Elvanidou, M.; Katifori, E.; Kaskanioti, M.; Soetens, S.; Kalpaxis, Th.; Bassiakos, Y.; Athanassas, C.; Hayden, B., & Brennan, T., (2007a) A multi-disciplinary approach to industrial sites of the Vrokastro region of Mirabello, eastern Crete. In: Proc. XXXIII Computer Applications and Quantitative Methods in Archaeology Conf. (Tomar, Portugal, March 2005): The World in Your Eyes ed A Figueiredo and G Velho CAA Portugal—Associacao para o Desenvolvimento das Aplicacoes, *Informaticas e Novas Technologiasem Arqueologia: Tomar,* 211–217.

[35] Sarris, A.; Dunn, R. K.; Rife, J. L.; Papadopoulos, N.; Kokkinou, E.; & Mundigler, C. (2007b) Geological and Geophysical investigations in the Roman cemetery at Kenchrai (Korinthia), Greece. *Archaelogical Prospection,* 14, 1-23.

[36] Sarris, A.; Papadopoulos, N.; Trigkas, V.; Kokkinou, E.; Alexakis, D.; Dimitrides, D.; Kakoulaki, G.; de Marco, E.; Seferou, E.; Aresti, G.; Shen, G.; Kondili, F.; Karaoulis, M.; Karifori, M.; Simiridanis, K.; Koustas, G.; Nikas-Karayiannis, Y.; Dogan, M.; Stamatis, G.; Kappa, E.; Lolos, Y.; & Kalpaxis, Th. (2007c) *Recovering the Urban Network of Ancient Sikyon Through Multi-component Geophysical Approaches*, CAA, 6p.

[37] Sarris, A.; Teichmann, M.; Seferou, P.; & Kokkinou E. (2010) Investigation of the Urban-Suburban Center of Ancient Nikopolis (Greece) through the Employment of Geophysical Prospection Techniques. In CAA2010, 5p.

[38] Papadopoulos, N. G.; Sarris, A.; Salvi, C. M.; Dederix S.; Soupios P.; & Dikmen U. (2012) Rediscovering the Small Theatre and Amphitheatre of Ancient Ierapytna (SE Crete) by Integrated Geophysical Methods. *Journal of Archaeological Science*, 39, 1960-1973.

[39] Sarris, A.; Papadopoulos, N.; Agapiou, A.; Salvi, M. C.; Hadjimitsis, D. G.; Parkinson W. A.; Yerkes R. W.; Gyucha, A.; & Duffy P. R. (2013) Integration of geophysical surveys, ground hyperspectral measurements, aerial and satellite imagery for archaeological prospection of prehistoric sites: the case study of Vészto-Mágor Tell, Hungary. *Journal of Archaeological Science*, 40, 1454-1470.

[40] Gorokhovich, Y.; Papadopoulos, N.; Soupios P.; & Barsukov P. (2014) Integrating GIS terrain modeling with geophysical surveys for archaeological investigation in Aptera, Crete, Special Issue "Advanced GIS terrain analysis for geophysical applications." *El Boletín Geológico y Minero, Spain,* 125 (3), 273-284.

[41] Barnett, C. T. (1976) Theoretical modeling of the magnetic and gravitational fields of an arbitrarily shaped three–dimensional body. *Geophysics*, 41(6), 1353–1364.

[42] Blakely, R. J.; & Simpson, R. W. (1986) Approximating edges of source bodies from magnetic or gravity anomalies. *Geophysics,* 51(7), 1494–1498.

[43] Blakely, R. J. (1996) Potential Theory in Gravity & Magnetic Applications; Cambridge University Press, 411p.

[44] Steeples, D. W.; Knapp, R. W.; & McElwee, C. D. (1986) Seismic reflection investigations of sinkholes beneath Interstate Highway 70 in Kansas. *Geophysics*, 51, 295-301.

[45] Miller, R. D.; Pullan, S. E.; Keiswetter, D. A.; Steeples, D. W; & Hunter, J. A. (1994) Field comparison of shallow P-wave seismic sources near Houston, Texas. *Geophysics,* 59, 1713-1728.

[46] Kokinou, E.; Kamberis, E.; Vafidis, A.; Monopolis, D.; Ananiadis, G.; & Zelelidis, A. (2005) Deep seismic reflection data from offshore Western Greece: a new crustal model for the Ionian sea. *Journal of Petroleum Geology,* 28 (2), 81-98.

[47] Hammer, S. (1939) Terrain corrections for gravimeter stations. *Geophysics*, 4, 184–194.

[48] Fajklewicz, Z. J. (1956) Underground gravity measurements as applied to coal mining. Results of investigations in Miechowice Colliery. *Archiwum Goimictwa*, I(4), 345–355.

[49] Fajklewicz, Z. J. (1976) Gravity Vertical Gradient Measurements for the Detection of Small Geologic and Anthropogenic Forms. *Geophysics,* 41(5), 1016.

[50] Butt, M. H. P. (1960) The use of rapid digital computing methods for direct gravity interpretation of sedimentary basins. *Geophysical Journal of the Royal Astronomical Society,* 3, 63–67.

[51] Colley, G. C. (1963) The detection of caves by gravity measurements. Geophysical Prospecting, 11 (1), 1–9.

[52] Arzi, A. A. (1975) Microgravity for engineering applications. *Geophysical Prospecting*, 23, 40825.

[53] Butler, D. K. (1984) Microgravimetric and gravity gradient techniques for detection of subsurface cavities. *Geophysics,* 49 (7), 1084–1096.

[54] Butler, D. K. (1995) Generalized gravity gradient analysis for 2–D inversion. *Geophysics*, 60(4), 1018–1028.

[55] Bear, G. W.; Al–Shukri, H. J; & Rudman, A. J. (1995) Linear inversion of gravity data for 3–D density distributions. *Geophysics*, 60(5), 1354–1364.

[56] Bishop, I.; Styles, P.; Emsley, S. J; & Ferguson, N. S. (1997) The Detection of Cavities Using the Microgravity Technique: Case Histories From Mining and Karstic Environments. In: Modern Geophysics in Engineering Geology; McCann, Fenning& Reeves, Eds.; *Geol. Soc. of London Special Publication*, 12, 155–168.

[57] Crawford, N. C.; Lewis, M. A.; Winter, S. A.; & Webster, J. A. (1999) Microgravity techniques for subsurface investigations of sinkhole collapses and for detection of groundwater flow paths through karst aquifers. In: Hydrology and Engineering Geology of Sinkholes and Karst; Seventh Multidisciplinary Conference on Sinkholes and the Engineering and Environmental Impacts of Karst; Beck, Petit & Herring, Eds.: Rotterdam, Harrisburg, Pennsylvania.

[58] McGrath, R. J.; Styles, P.; Thomas, E. T.; & Neale, S. (2002) Integrated high resolution geophysical investigations as potential tools for water resource investigations in karst terrain. *Environmental Geology*, 42, 552–557.

[59] Branston, M. W.; & Styles, P. (2003) The application of time–lapse microgravity for the investigation and monitoring of subsidence at Northwich, Cheshire. *Quarterly Journal of Eng Geology & Hydrogeology*, 36(3), 231–244.

[60] Branston, M. W.; & Styles, P. (2006) Site characterization and assessment using the microgravity technique: a case history. *Near Surface Geophysics*, 4(6), 377–385.

[61] Pdyke, Geophysical Software Solutions, http://www.geoss.com.au/downloads.html.

[62] Pirttijärvi, M. (2014) Grablox2, Gravity inversion using block model 2.

[63] Nabighian, M. N.; Ander, M. E.; Grauch, V. J. S.; Hansen, R. O.; LaFehr, T. R.; Li, Y.; Pearson, W. C.; Peirce, J. W.; Phillips, J. D.; & Ruder, M. E. (2005) The historical development of the gravity method in exploration, *Geophysics*, 70(6), 63ND-89ND.

[64] Thompson, T.; & Oldfield, F. Environmental Magnetism; Allen and Unwin: London, 1986.

[65] Strzyszcz, Z.; Magiera, T.; & Heller, F. (1996) The influence of industrial emissions on the magnetic susceptibility of soils in Upper Silesia. *Studia Geophysica et Geodaetica*, 40, 276–286.

[66] Petrovský, E.; & Ellwood, B. B. (1999) Magnetic Monitoring of Air-, Land- and Water Pollution. In: B. A. Maher and R. Thompson, Eds, Quaternary Climates, Environments and Magnetism; Cambridge University Press, Cambridge: U.K., pp 279–322.

[67] Kapička, A.; Jordanova, N.; Petrovský, E.; & Podrázský, V. (2003) Magnetic study of weakly contaminated forest soils. *Water, Air, & Soil Pollution,* 148, 31−44.

[68] Kokinou, E. 2010 Magnetic properties of soils around a local power plant (Chania, Crete, Greece). In: *Proc. of the XIX Congress of the Carpathian Balkan Geological Association*, Special Volume, 100, pp 594-600.

[69] Kaufman, A.; & Keller, G. (1983) Frequency and transient soundings; Methods in geochemistry and geophysics; Elsevier: Amsterdam.

[70] Fitterman, D. V.; & Stewart, M. T. (1986) Transient electromagnetic sounding for groundwater. *Geophysics*, 51, 995–1005.

[71] Stewart, M.; & Gay, M. C. (1986) Evaluation of transient electromagnetic soundings for deep detection of conductive fluids. *Ground Water*, 24, 351–356.

[72] Goldman, M.; Arad, A.; Kafri, U.; Gilad, D.; Melloul. A. (1988) Detection of freshwater/seawater interface by the time domain electromagnetic (TDEM) method in Israel. *Naturwet Tijdsehr*, 70, 339–344.

[73] Mills, T.; Hoekstra, P.; Blohm, M.; & Evans, L. (1998) Time domain electromagnetic soundings for mapping sea-water intrusion in Monterey County, California. *Ground Water*, 26, 771–782.

[74] Nabighian, M. N.; Macnae, J. C. (1991) Time domain electromagnetic prospecting methods. In: Electromagnetic methods in applied geophysics: Tulsa, pp 427–520.

[75] McNeill, J. D. (1994) Principles and applications of time domain electromagnetic techniques for resistivity sounding. *Geonics Ltd. Technical Note* TN-27.

[76] Kanta, A.; Soupios, P.; Vallianatos, F.; Rust, D.; & Barsukov, P. (2009) The application of time domain electromagnetic method to characterize the Keritis Basin, in Western Crete, Greece; 3rd IASME/WSEAS international conference on geology and seismology (GES '09): Cambridge.

[77] Reynolds, J. M. An introduction to applied and environmental geophysics; Willey: New York, 2010; ISBN-13:9780471485353.

[78] Soupios, P; Kalisperi, D.; Kanta, A.; Kouli, M.; Barsukov, P., & Vallianatos, F. (2010) Coastal aquifer assessment based on geological and geophysical survey, North Western Crete, Greece. *Environ earth Sci,* 61(1), 63–77.

[79] Barsukov, P. O; Fainberg, E. B; & Khabensky, E. O. (2007) Shallow investigation by TEM-FAST technique: methodology and case histories. In: Methods of geochemistry and geophysics; Spichak, V. V.; Ed.; Elsevier: Amsterdam, pp 55–77.

[80] Koutsios, A; Kontopoulos, N; Kalisperi, D.; Soupios, P.; & Avramidis, P. (2010) Sedimentological and geophysical observations in the delta plain of Selinous river: The lost city of ancient Helike. In: Proc. of the 12th international congress, *Bulletin of the Geological Society of Greece,* 4, pp 654-662.

[81] Annan, A. P. Ground Penetrating Radar; Workshop Notes; Sensors And Software Inc.: Canada, 1992.

[82] Mellett, J. S. (1995) Ground penetrating radar applications in engineering, environmental management, and geology. *Journal of Applied Geophysics*, 33(1–3), 157–66.

[83] Goodman, D.; Nishimura, Y.; & Rogers, J. D. (1995) GPR time slices in archaeological prospection. *Archaeological prospection*, 2(2), 85–89.

[84] Matthews, S. L. (1998) Application of Subsurface Radar as an Investigative Technique; BRE Reports, 340. IHS BRE Press.

[85] Massonnet, D.; & Feigl, K. L. (1998) Radar interferometry and its application to changes in the earth's surface. *Reviews of Geophysics*, 36, 441–500.

[86] Conyers, L. B. (2010) Ground–penetrating radar for anthropological research. *Antiquity*, 84 (323), 175–184.

[87] Talwani, M.; & Ewing, (1960) M. Rapid Computation of gravitational attraction of three–dimensional bodies of arbitrary shape. *Geophysics,* 25, 203–25.

[88] Plouff, D. (1976) Gravity and magnetic fields of polygonal prisms and application to magnetic. *Geophysics*, 41(4), 727–41.

[89] Won, I.; & Bevis, M. (1987) Computing the gravitational and magnetic anomalies due to a polygon–algorithms and FORTRAN subroutines. *Geophysics*, 52(2), 232–38.

[90] Singh, B.; &Guptasarma, D. (2001) New method for fast computation of gravity and magnetic anomalies from arbitrary polyhedra. *Geophysics,* 66(8), 521–26.

[91] Thompson, D. T. (1982) EULDPH: A new technique for making computer–assisted depth estimates from magnetic data. *Geophysics,* 47(1), 31–37.

[92] Tarantola, A. Inverse Problem Theory; Elsevier: Amsterdam, 1987.

[93] Li, Y.; & Oldenburg, D. W. (1996) 3–D inversion of magnetic data. *Geophysics*, 1996, 61(2), 394–408.

[94] Li, Y.; & Oldenburg, D. W. (1998) 3–D inversion of gravity data. *Geophysics*, 1998, 63(1), 109–19.

[95] Gallardo–Delgado, L. A.; Pérez–Flores, M. A.; & Gómez–Treviño, E. (2003) A versatile algorithm for joint 3D inversion of gravity and magnetic data. *Geophysics,* 68 (3), 949–959.

[96] Hsu, S. K.; Sibuet, J. C.; & Shyu, C. T. (1996) High–resolution detection of geological boundaries from potential–field anomalies: An enhanced analytic signal technique. *Geophysics*, 61(2), 373–86.

[97] Jol, H. M; & Bristow, C. S. (2003) GPR in sediments: advice on data collection, basic processing and interpretation, a good practice guide. Geological Society, London, *Special Publications*, 211, 9–27.

[98] Kokinou, E.; Panagiotakis, C.; & Sarris, A. (2010) Application of Skeletonization on Geophysical Images. In: Proc. of the XIX Congress of the Carpathian Balkan Geological Association, Special Volume, 99, pp 338-343.

[99] Panagiotakis, C.; Kokinou, E.; & Sarris, A. (2012) Curvilinear Structure Enhancement and Detection in Geophysical images Based on a Multiple Filtering Scheme. *IEEE Transactions on Geoscience and Remote Sensing,* 49, 2040-2048.

[100] Panagiotakis, C.; & Kokinou, E. (2015) Linear Pattern Detection of Geological Faults via a Topology and Shape Optimization Method, *IEEE Journal of Selected Topics in Applied Earth Observations and Remote Sensing*, 8(1), 3-11.

[101] Doll, W.; Rohdewald, S.; Sheehan J.; Burton B. (2010) Processing of Seismic Refraction Tomography Data; Short course notes; *Symposium on the Application of Geophysics to Engineering and Environmental Problems*, SAGEEP, 83p.

[102] Murray, A. S.; & Tracey, R. M. (2003) Basic practice in gravity surveying; Australian Geological Survey Organization (AGSO): htpp://www.agso.gov.au/.

[103] Kanta, A.; Soupios, P.; Barsukov, P.; Kouli, M., & Vallianatos, F. (2013) Aquifer characterization using shallow geophysics in the Keritis Basin of Western Crete, Greece. *Environmental Earth Sciences*, 70(5), 2153-2165.

[104] Kyriacou, A.; Lasaridi, K. E.; Kotsou, M.; Balis, C.; & Pilidis G. (2005) Combined bioremediation and advanced oxidation of green table olive processing wastewater, *Clay Miner*, 40, 1401–1408.

[105] Kirmizakis, P., Soupios, P., Simyrdanis, K., Kirkou, S., Papadopoulos, N., Tsourlos P., Ntarlagiannis D., Robinson, J., Slater, L. D,& and Kim J.-H. (2015) Geoelectrical characterization of an olive oil mill waste (OOMW) site; SAGEEP 2015: Texas Austin, USA.

[106] Kavvadias, V.; Doula, M. K.; Komnitsas, K.; Liakopoulou, N. (2010) Disposal of olive oil mill wastes in evaporation ponds: Effects on soil properties. *Journal of Hazardous Materials*, 182, 144–155.

[107] Richards, K.; Revil, A.; Jardani, A.; Henderson, F.; Batzle, M.; & Haas, A. (2010) Pattern of shallow ground water flow at Mount Princeton Hot Springs, Colorado, using geoelectrical methods. *Journal of Volcanology and Geothermal Research*, 198, 217-232.

[108] Mac Donald, A. M.; Calow, R. C.; Nicol, A.; Hope, B.; & Robins, N. S. (2001) Ethiopia: water security and drought; British Geological Survey, Technical Report WC/01/02.

[109] Meju, M. A; Fontes, S. L.; Lima, J. P.; Ulugergerli, E. U.; & Carrasquilla, A. A. (1999) Regional aquifer mapping using combined VES-TEM-AMT/EMAP methods in the semi-arid eastern margin of Parnaiba Basin, Brazil. *Geophysics*, 64, 337–356.

[110] Peavy, S. T.; & Valentino, D. W. (1999) Variability of electrical resistivity at the Rice Creek Field Station, Oswego, New York: Implications for the distribution of groundwater. In: Proceedings for SAGEEP: Oakland, CA, pp 349–358.

[111] Sørensen, K. I; & Søndergaard, V. (1999) H Large-scale geophysical mapping and its application for ground water protection in urban areas. In: Proceedings for SAGEEP, Oakland, CA, pp 481–486.

[112] Gwaze, P.; Dahlin, T.; Owen, R.; Gwavava, O.; & Danielsen, J. (2000) Geophysical investigations of the Karoo Aquifer at Nyamandhlovu, Zimbabwe. In: Proceedings for the EEGS-ES, annual meeting, Bochum, Germany.

[113] Corriols, M.; Bjelm, L.; & Dahlin, T. (2000) Resistivity surveying applied to groundwater studies in the Leon-Chinandega Plains, Nicaragua, Central America. Proceedings for the EEGS-ES, annual meeting, Bochum, Germany.

[114] Mendoza, A.; Bjelm, L.; & Dahlin, T. (2000) Resistivity imaging as a tool for groundwater studies at Santo Domingo, Central Nicaragua. In: Proceedings for the EEGS-ES, annual meeting: Bochum, Germany.

[115] Miele, M. J.; Jansen, J.; Arizpe, J. E. D.; Magallanes-Mercado, M. A. (2000) A regional groundwater evaluation using magnetotelluric sounds for Monclova, Mexico. In: Proceedings for SAGEE: Washington, DC, pp 699–708.

[116] Maillol, J. M.; Ortega-Ramirez, J.; Bandy, W. L; & Valiente-Banuet, A. (2000) Contribution of electrical resistivity methods to paleoenvironmental reconstruction and groundwater exploration in the Chihuahua Desert, Mexico. In: Proceedings for SAGEEP: Washington, DC, pp 1067–1076.

[117] Wynn, J.; Pool, D.; Bultman, M.; Gettings, M., & Lemieux, J. (2000) Airborne EM as a 3D aquifer mapping tool. In: Proceedings for SAGEEP 2000: Washington, DC, pp 93–100.

[118] Farrell, D. A.; La Femina. P.; Armstrong, A.; Sandberg, S.; & Rogers N. (2000) Constraining hydrogeologic models using geophysical technique: Case study Fortymile Wash and Amargosa Desert, Southern Nevada. In: Proceedings for SAGEE: Washington, DC, pp 213–222.

[119] Paine, J. G.; Angle, E. S; & Petrossian, R. (2000) Identification and assessing groundwater in the lower Rio Grande Valley, Texas, using airborne electromagnetic induction. In: Proceedings for SAGEEP: Washington, DC, pp 83–92.

[120] Soupios, P.; Kouli, M.; Vallianatos, F.; Vafidis, A; & Stavroulakis; G. (2007d) Estimation of Aquifer Parameters from Surficial Geophysical Methods: A Case Study of Keritis Basin in Crete. *Journal of Hydrology*, 338, 122-131.

[121] Soupios, P.; Kalisperi, D.; Kanta, A.; Kouli, M.; Barsukov, P.; & Vallianatos, F. (2009) Coastal aquifer assessment based on geological and geophysical survey, North Western Crete, Greece. *Environmental Earth Sciences*, 61, 63–77.

[122] Klimis, N. S.; Papazachos, C. B.; & Efremidis, Ch. F. (1999) Determination of the behavior of a sedimentary rock mass: comparison of measured static and dynamic properties; In: Proc. 9th Int. Conf. on Rock Mechanics: Paris, France.

[123] Luna, R.; & Jadi, H. (2000) Determination of dynamic soil properties using geophysical methods. In: Proc. 1st Int. Conf. on the Application of Geophysical and NDT Methodologies to Transportation Facilities and Infrastructure Geophysics; Federal Highway Administration, Saint Louis, MO, pp 1–15.

[124] Othman, A. A. (2005) Construed geotechnical characteristics of foundation beds by seismic measurements. *J Geophys Eng*, 2 126–38.

[125] Savvaidis, A; Tsokas, G.; Soupios, P; Vargemezis, G.; Manakou, M.; Tsourlos, P., & Fikos, I. (1999) Geophysical prospecting in the Krousovitis dam (N Greece) by seismic and resistivity geophysical methods. *J Balkan GeophysSoc*, 2, 128–39.

[126] Savvaidis, A.; Margaris, B.; Theodoulidis, N.; Lekidis, V.; Karakostas, Ch.; Loupasakis, C.; Rozos D.; Soupios, P.; Mangriotis, M-D.; Dikmen, U.; Tsangaratos, P.; Kokinou, E.; Vafidis, A.; Rondoyanni, Th.; Kalogeras, I.; Koutrakis, S.; Sarris, A.; & Papadopoulos, N. (2014) Geo-Characterization at selected accelerometric stations in Crete (Greece) and comparison of earthquake data recordings with EC8 elastic spectra. *C Europ J of Geosciences*, 6 (1), 88-103.

[127] Soupios, P. M.; Papazachos, C. B.; Vargemezis, G.; & Fikos, I. (2005) Application of modern seismic methods for geotechnical site characterization. In: Proc. Int. Workshop in Geoenvironment and Geotechnics: Milos Island, Greece, pp 163–70.

[128] Soupios, P.; Papazachos, C. B.; Vargemezis, G.; Savvaidis, A.; (2006) In situ geophysical investigation to evaluate dynamic soil properties at the Ilarionas Dam, Northern Greece. In: Proc. 2nd Int. Conf. Advances in Mineral Resources Management and Environmental Geotechnology; Heliotopos Conferences: Hania, Crete, Greece, pp 149–56.

[129] Soupios, P.; Loupasakis, C.; & Vallianatos, F. (2008) Reconstructing former urban environments by combining geophysical electrical methods and geotechnical engineering – an example from Chania, Greece. *J Geophys Eng*, 186-194.

[130] Venkateswara, V. R.; Srinivasa, R.; Prakasa, R. B. S.; &Koteswara, R. P. (2004) Bedrock investigation by seismic refraction method—a case study *J. Ind. Geophys. Union*, 8, 223–8.

[131] Soupios, P.; Georgakopoulos, P.; Papadopoulos, N.; Vallianatos, F.; Sarris, A.; & Makris, J. P. (2007c) Use of Engineering Geophysics to Investigate a Site for a Building Foundation, *J GeophysEng*, 4, 94–103.

[132] Loke, M. H.; & Barker, R. D. (1996) Practical techniques for 3D resistivity surveys and data inversion. *Geophysical Prospection*, 44, 499–523.

[133] Loke, M. H. (1997) Rapid 2D resistivity inversion using the least-squares method RES2DINV Program; Manual: Penang, Malaysia.

[134] Loke, M. H.; Chambers, J. E.; Rucker, D. F.; Kuras, O.; & Wilkinson P. B. (2014) Recent developments in the direct-current geoelectrical imaging method. *Journal of Applied Geophysics*, 95, 135-156.

[135] Shahrukh, M.; Soupios, P.; Papadopoulos; N.; & Sarris, A. (2012) Geophysical investigations at the Istron archaeological site, eastern Crete, Greece using 3D Refraction Seismic and Geoelectrical Tomography. *J Geophys Eng*, 9, 749–760.

[136] Hayden, B. et al. A multidisciplinary approach to the archaeoenvironment and the cultural residues in Istron, Mirabelleo Gulf; 5th Symp. of Greek Archaeometric Society; Papazisi (in Greek), 2000, pp 327–44.

[137] Hayden, B. J.; Dierckx, H.; Harrison, G. W. M.; Moody, J.; Postma, G.; Rackham, O.; & Stallsmith, A. B. (2004) Reports on the Vrokastro Area, Eastern Crete, The Settlement History of the Vrokastro Area and Related Studies; University Museum Monograph; University of Pennsylvania: Philadelphia, PA, 2.

[138] Sarris, A.; Kokkinou, E.; Kouriati, K.; Aedona, E.; Karagianni, L.; Vargemezis, G.; Stamatis, G.; Elvanidou, M.; Katifori, E.; Kaskanioti, M.; Soetens, S.; Kalpaxis, Th.; Bassiakos, Y.; Athanassas, C.; Hayden, B.; Brennan, T. (2005) Geophysical and geomorphological studies at the wider Istron area. In: 6th Int. Conf. on Archaeological Prospection National Research Council (CNR) (Rome), pp 332–5.

[139] Zacharias, N.; Bassiakos; Y; Hayden, B.; Theodorakopoulou, K.; & Michael, T. C. (2009) Luminescence dating of deltaic deposits from eastern Crete, Greece: geoarchaeological implications. *Geomorphology,* 109 46–53.

Reviewed by Prof. Francesco Sdao, School of Engineering, University of Basilicata, Italy; francesco.sdao@unibas.it.

Asc. Prof. George Vargemezis, School of Geology, Aristotle University of Thessaloniki, Greece; varge@geo.auth.gr.

In: Geophysics
Editor: Gemma Aiello

ISBN: 978-1-63484-831-2

Chapter 2

APPLICATION OF INNOVATIVE GEOPHYSICAL TECHNIQUES IN COASTAL AREAS

Vincenzo Di Fiore*[*]*, Michele Punzo, Daniela Tarallo and Giuseppe Cavuoto

Institute for Coastal Marine Environment (IAMC), Italian National Research Council (CNR), Calata Porta di Massa - Porto di Napoli, Naples, Italy

ABSTRACT

Coastal and transitional areas represent highly dynamic, quickly-changing environments. These areas are subject to multiple interactions with the marine-land-fluvial systems processes. Different types of processes such as coastal erosion, seawater intrusion, pollution and transport of sediment affect these areas. The geophysical surveys permit to improve the understanding of these phenomena and consequently to assess the vulnerability and the potential of the coastal zone. The development of geophysical models integrated into management coastal system is the key to achieve reliable prediction model. Moreover, coastal areas are among the most populated and urbanized areas. For these reasons, the application of geophysical minimally invasive methods, such as seismic and geoelectric techniques, are the most appropriate tool to provide geology and to collect data on land-sea zone.

In coastal areas, the seismic reflection/refraction survey plays a key role to increase the knowledge of civil engineers. In particular, seismic survey provide information on subsoil geometry and evaluate the hydraulic and mechanical characteristics necessary in the planning, design and construction stages. The seismic method, routinely employed to obtain information on subsurface geology, has a limited use in the transitional zone. This chapter describes a non standard approach to acquire seismic data in the backshore-foreshore area, joining up land and marine seismic data, using a marine streamer connected to a land cable. This method allows to obtain a seismic site characterization in terms of primary wave velocities and shear wave velocities (both the P-wave and S-wave velocities are used to define the elastic modules and geotechnical parameters), to identify major acoustic impedances within the overburden (e.g., bedrock distribution), to delineate

[*] Corresponding Author address: e-mail: vincenzo.difiore@cnr.it.

any local subsurface structure (e.g., existence of weak zones such as faulted areas or weathering zone).

The monitoring of the intrusion of seawater also represents a priority for the safeguard of coastal areas. The identification of changes in the freshwater-saltwater interface position can be a useful element for the rationalization of water resources in order to guide the choice of use of these areas. To address this issue, we use electrical surveys, in particular 3D Electrical Resistivity Tomography, to monitor and image fresh-salt water transition.

Keywords: coastal areas, geophysical techniques, seismic data, Electrical Resistivity Tomography

INTRODUCTION

According to the Integrated Coastal Zone Management (ICZM) of the European Commission [1], coastal areas are of great environmental, economic, social and cultural relevance. Therefore, the implementation of suitable monitoring and protection actions is fundamental for their preservation in order to ensure the future use of this resource. Such actions are based on an ecosystem perspective to preserve the integrity and functioning of the coastal environment and to plan sustainable resource management of both the marine and terrestrial components. The planning and the management of natural resources has as objective the promotion of economic and social welfare of coastal zones.

The knowledge of the geological and environmental aspects of coastal area is extremely important for the evaluation of hydrogeological and geomechanical subsurface characteristics. These environments are characterized by multiple processes such as erosion, sediment transport, water-sea intrusion and pollution. The definition of these dynamics enables the management and the correct planning of coastal areas. Because of the highly urbanized areas it is needed to use minimally invasive methods.

For these reasons, the application of geophysical methods, such as seismic and geoelectric techniques, is the most appropriate to track underground geology and to collect data on land-sea transitional zone.

Reflection and refraction seismic methods, easily performed in the land or sea areas, are difficult to apply in transitional areas. To exceed this limit, we developed a nonstandard approach to acquire seismic data in the backshore-foreshore area, joining up land and marine seismic data, using a marine streamer connected to land cable [2]. The use of seismic encounters several difficulties when used in shallow water areas: the water column is too shallow for efficient towed streamer operations and too congested for shooting with conventional marine seismic source [3]. The primary limitation is the strong signal reverberation that increases the background noise level. Hence, obtaining a continuous profile of the subsurface in the land-sea transition zone can be a challenging task. The use of this type of acquisition allowed to obtain a seismic characterization of the area in terms of P and S velocity waves vs depth. This characterization has the aim of increasing the awareness of civil engineering, especially to obtain information on surface condition and to evaluate the mechanical characteristics of the subsoil necessary in the planning, design and construction stages.

Another common problem in the coastal area is the saltwater intrusion, induced by the flow of seawater into freshwater aquifers. The salinization vulnerability is probably the most common and diffuse problem in a costal aquifer. The boundary between salt water and fresh water is not very clear; the transitional zone are brackish, due to the mixing of salt water and fresh water. Under normal conditions, fresh water flow from land aquifers up to sea. Generally, groundwater moves from area with higher hydraulic gradient to areas with lower hydraulic gradient [4]. Among the geophysical methods applied for the location and movement of saltwater intrusion, best results were obtained through Electrical Resistivity Tomography (ERT) [5, 6]. Many hydrological processes can be expected to provide significant contrasts in resistivity, consequently, ERT has been adopted as a tool for new researches within the field of the hydrology. Several studies have demonstrated the ability of ERT to visualize the hydrogeological structures within laboratory cores [7, 8], to monitor fluid or contaminant migration at the field scale [9, 10, 11] and to ascertain the efficiency of new contaminant remediation processes [12, 13].

We applied 3D ERT to assess changes in the freshwater-brine interface in both space and time. This acquisition has allowed to obtain a 4D monitoring salt wedge in a volumetric specific subsurface.

METHODS

Joining up Land and Marine Seismic Data

The innovative seismic method consists in the acquisition of a continuous seismic profile in the backshore-foreshore area, joining up land and marine seismic data. This method permitted the recording of high-resolution data in a transitional zone characterized by a complex surface geology. In this way, it was possible to define both compressional velocities (Vp velocities derived from tomography of refracted first arrivals) and shear wave velocities (Vs velocities derived from inversion of the phase velocities dispersion curves). The definition of the P-wave velocity (Vp) is important in order to provide information on the geometry of the subsurface geological formations (e.g., bedrock distribution) and to locate possible stratigraphic/tectonic features (e.g., structural boundaries, active faults, buried small-scale structures, weak zones such as faulted areas or weathering areas). Instead, the Vs is an essential parameter, widely used in order to obtain a detailed characterization of soil properties and to predict amplification of the seismic ground motion for earthquake engineering design purposes. Vs are empirically used to evaluate the elastic modules and the geotechnical parameters (within Vp) and to identify major contrasts of acoustic impedance within the overburden strata. This has provided useful insights to map the 2D bedrock surface and to evaluate the ground stiffness and the dynamic behavior of the soils.

The acquisition has been carried out using the same seismic survey design for the seismic refraction 2D tomography and for the analysis of surface waves. Land and marine 10 Hz sensors have been used during the seismic acquisition. The hydrophones have been located from the shoreline to the sea bottom, while the geophones have been located on the beach. Sea-bottom cables have been connected to land cables to record a continuous sea-land seismic profile. The main idea, in fact, was to acquire seismic refraction and MASW datasets using one single field operation. In our particular case, we acquired simultaneously two sub-dataset: land-land and land-marine, since shooting from the offshore side was unfeasible. To achieve symmetrical land-marine ray coverage, the seismic sources were placed in the middle of the total recording array (between the land and the marine receivers respectively).

This method demonstrates that land-water bottom refraction survey is a very efficient and low cost method to map the 2D subsurface features in transitional zones. The accurate mapping of the substratum in the water-covered sectors is one of the most crucial factors in the construction of harbour settlements and bridges in coastal areas. Moreover, the obtained information is useful in a well-defined beach protection strategy (e.g., a volume estimation of the sediment for the purpose of management of bottom sediments). In general, a fast and reliable method for mapping transitional zones may contribute to the measurement of geotechnical properties for applications such as: engineering and construction, coastal management, recreational reformation as well as scientific approaches like coastal evolution researches and paleo-environmental studies.

It is extremely difficult to obtain a high-resolution seismic data in a marine transition area; few cases refer to "really" high-resolution land-marine seismic acquisition. Recently, for example, Stucchi et al. 2005 [14] carried out a joint land-marine seismic survey to characterize the body of a huge landslide; Kaufmann et al. 2005 [15] acquired seismic data in a shallow marine environment and on adjacent land. We present two seismic in-situ tests to produce high-resolution models of the upper 30 meters of the subsurface, through the integration of 2D seismic refraction tomography with MASW in presence of a shallow bedrock [2].

We have tested the performance of our methodology choosing two transitional areas of limited extension from shore to foreshore with very low nearshore bathymetric gradient. In particular, we have selected two rocky-sandy beaches: the first one is positioned in the Procida Island, located in the Phlegrean Fields volcanic district (Figure 1); the second one is placed in the Ventotene island, which belongs to the Eastern Pontine Archipelago (Figure1). As a general rule, the aim of both the surveys was to provide information on the geometry of the subsurface geological formations (0 – 30 meters) and to locate possible stratigraphic/tectonic features (e.g., structural boundaries, active faults, buried small-scale structures). The outcome of the survey in the Procida area was to obtain information on beach sediments and their seaward continuation in an area characterized by strong erosional trends [16].

For Ventotene test site, the main objective was to characterize the very irregular tuff deposit. For the latter, we used the stratigraphic information coming from three boreholes carried out by the local administration [17]. This information has been used in order to calibrate our geophysical model. We used for the acquisition the same seismic survey design for seismic refraction 2D tomography and MASW methods. Land-sea bottom seismic surveys consist of Procida SHR profile (or SHR1) which has a SW-NE strike and a length of 152.5 meters (Figure 2b) and Ventotene SHR profile (or SHR2) which moves with a NE-SW strike and a length of 142.5 meters (Figure 3b). Both profiles are characterized by a very low topographic relief; therefore we use a 2x vertical exaggeration to represent the small topographic gradient (Figures 2a and 3a).

During acquisition sea-bottom cables have been connected to land cables to form a continuous sea-land seismic profile: 24 receivers (10 Hz hydrophones) were located from the shoreline to the sea bottom until the depth of about 1 m, while the other 24 receivers (vertical

10 Hz geophones) were located on the beach. We used 12 shot points and two shots at each location were stacked to improve signal strength; the source moveout was 2.5 meters In particular, for Procida test site, we installed on the beach a second seismic array made of 24 P-wave 4.5 Hz geophones coincident with 10 Hz sensors. The sources were placed in the middle of the seismic array: between the geophones and hydrophones positions. The 48 vertical geophones/hydrophones were placed at 2.5 m intervals.

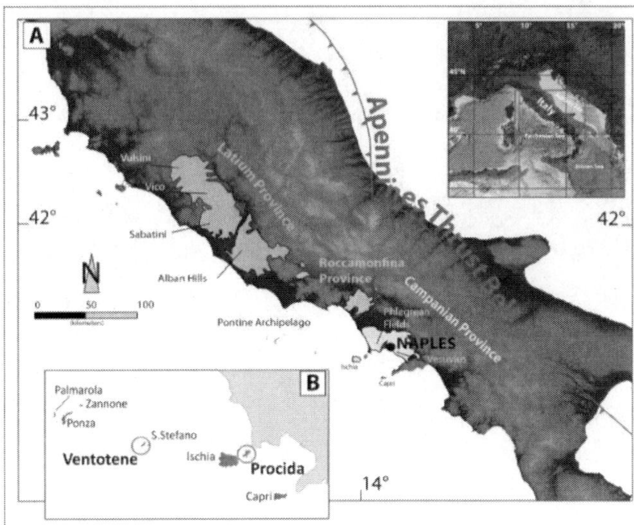

Figure 1. A) Geological sketch map of Roman Comagmatic Region including the Latium Province, the Roccamonfina Province, the Pontine Archipelago and the Campanian Province. B) Particular of the Pontine Archipelago and of the Phlegrean Fields volcanic zone. The red circles evidence the study areas: Ventotene and Procida islands (Di Fiore et al., 2015, modified).

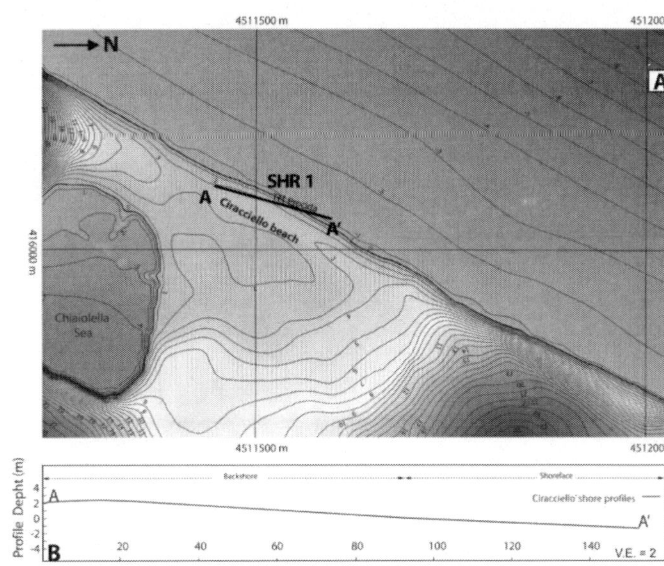

Figure 2. Procida land-marine profile. A) Backshore-foreshore topographic profile; B) Location of the SHR1 seismic line (Di Fiore et al., 2015, modified).

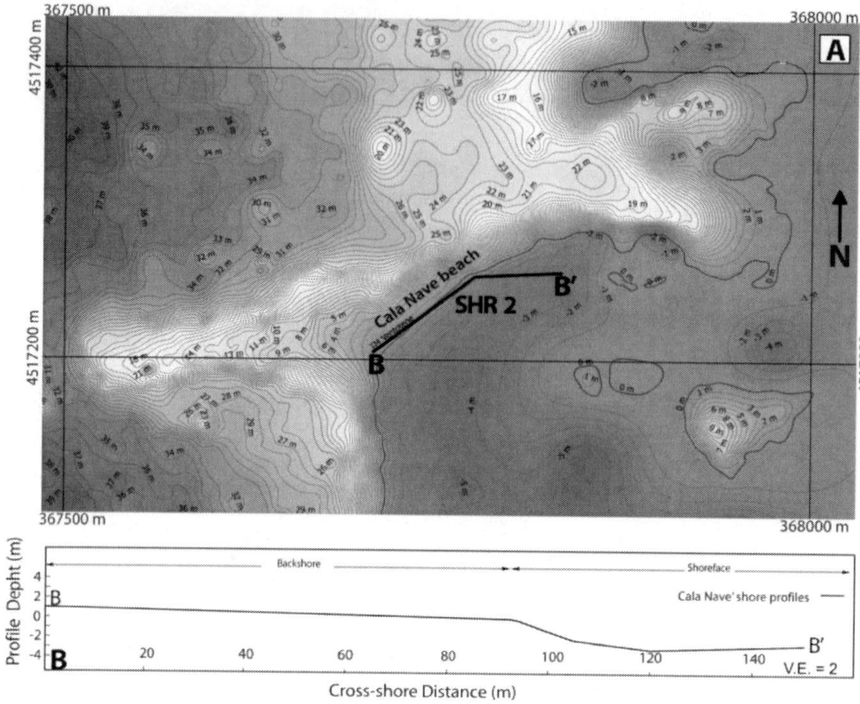

Figure 3. Trace of Ventotene profile. A) Backshore-foreshore topographic profile; B) Location of the SHR2 seismic line (Di Fiore et al., 2015, modified).

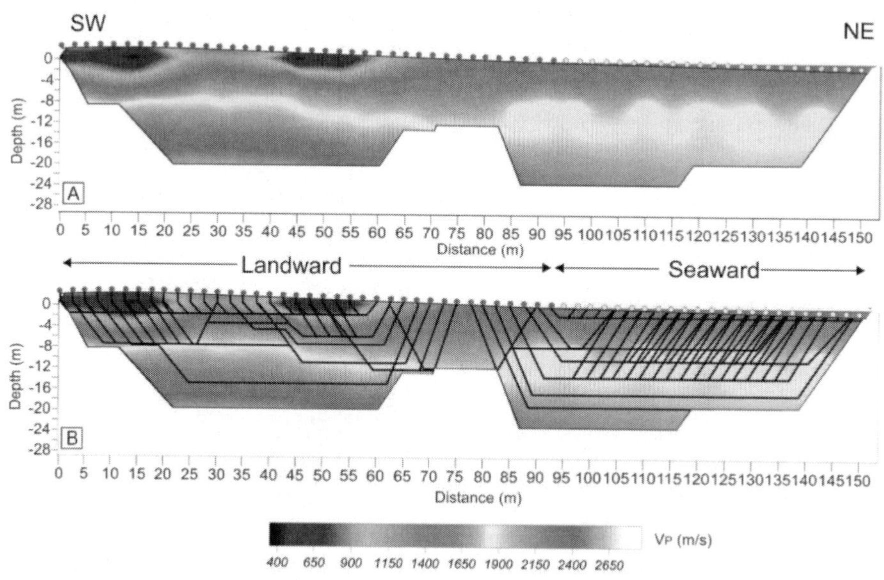

Figure 4. A) SHR1 final tomographic image. Velocity range is 400 to 2500 m/s. The green dots represent the vertical geophones, the yellow dots represent the hydrophones. The sources (red dots) were placed between the two seismic arrays. B) Ray density coverage diagram for SHR1 tomography. The region not sampled by ray paths is blank (Di Fiore et al., 2015, modified).

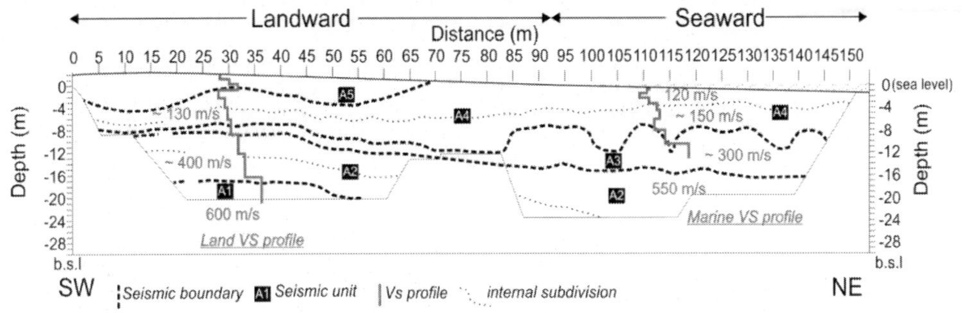

Figure 5. Line drawing of the SHR1 profile; the main seismic units were separated by dashed lines. The grey continuous lines show the projection of inland and marine VS profiles with the corresponding velocities values (Di Fiore et al., 2015, modified).

For Procida test site, the correlation between the tomographic model (Figure 4) and the related MASW dataset with the stratigraphic information (Figure 5) proposed by several authors [18] has suggested the following stratigraphic sequence, starting from the bottom:

- Unit A1 showing higher P-wave (> 2500 m/s) and S-wave (> 600 m/s) velocities is probably made of dense volcanic materials;
- Unit A2 with V_P field of about 2100-2400 m/s and S-wave velocities varying between 400 and 500 m/s, consists of volcanic materials;
- Unit A3 with P-wave velocities of about 1800-2100 m/s and S-wave velocities of about 300 m/s, is probably constituted by chaotic and massive pyroclastic deposits;
- Unit A4 is composed of saturated volcanic sand (V_P = 1100-1800 m/s); in the upper part it seems very loose showing low S-wave velocity (about 130 m/s);
- Unit A5 is formed by coarse-grained sandy beach deposits (V_P = 400-1100 m/s).

For Ventotene test site, the velocity models of seismic waves (Vp and Vs) have been classified by using borehole stratigraphic information (Figures 6 and 7). It was found that the soil stratification in the Cala Nave test area can be classified as sandy pyroclastic deposits overlying a more massive pyroclastic unit. The boundary between these two formations has been found at a depths between 10-15 meters.

The integration of geophysical and borehole information suggests the following stratigraphic scheme:

The first layer (U1), with a P-wave velocity exceeding 3000 m/s, consists of a matrix supported and massive pyroclastic unit (probably the PGT tuff deposits outcropping on the sea cliff);

Unit U2, with P-wave velocities about 2500 m/s and S-wave velocities about 750 m/s, is correlated with chaotic and massive pyroclastic flow deposits (ash matrix);

Units U3 (V_P = 1100-2500 m/s; V_S~700 m/s) and U4 (V_P~1000 m/s, V_S = 520 m/s) probably consist of reworked pyroclastic levels of nearshore shallow marine environment;

The shallow unit (U5) belongs to loose sand deposits.

The obtained results have demonstrated that the land-water bottom refraction survey is a very efficient and low cost method to map the 2D subsurface features in transitional areas. The accurate mapping of the substratum in the water-covered sectors is one of the most crucial factors in the construction of harbor settlements and bridges in the coastal areas.

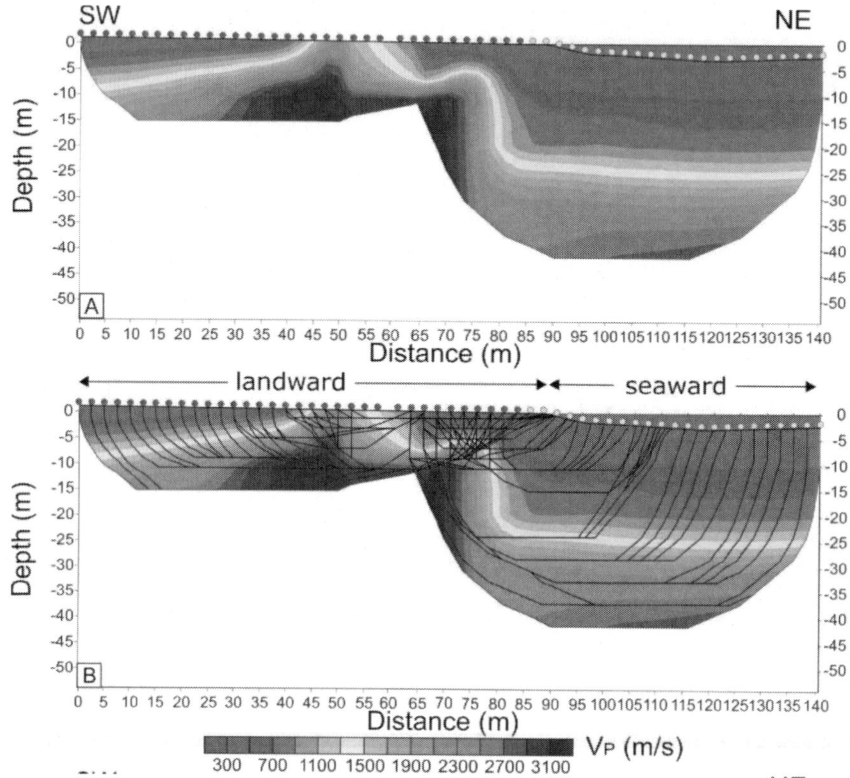

Figure 6. A) SHR2 tomographic image. The green dots represent the vertical geophones, the yellow dots represent the hydrophones. The sources (red dots) were placed between the two seismic arrays. As shown by the ray density (B), the maximum penetration depth is about 40 m with VP velocity ranging from 600 m/s to about 3200 m/s (Di Fiore et al., 2015, modified).

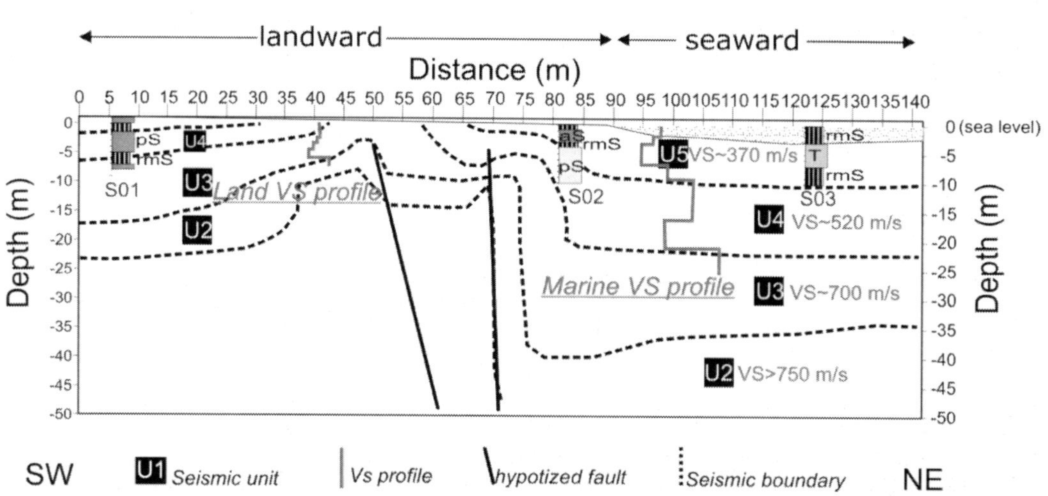

Figure 7. Line drawing of the SHR2 profile. The grey continuous lines point out the shallow Vs velocity estimated by the dispersion curves of the Rayleigh waves. The five seismic units are separated by dashed lines. The black solid lines denote possible faults. Boreholes log symbols: rmS: reworked marine Sand; aS: ash Sand; pS: pyroclastic Sand; T: Tuff (Di Fiore et al., 2015, modified).

4D Monitoring of Sea Water Intrusion by ERT

ERT method consists of the experimental determination of the apparent resistivity (ρ) of a given material through joint measurements of electric current intensity and voltage introduced into the subsoil with separate couples of electrodes, driven in the ground surface. All natural rocks can conduct electricity when subjected to an electric field. The measure of the ability of the rock to conduct electric current is known as dielectric constant which describes a material's capability (capacity) to hold a charge, which also measure the material's ability to polarize when subject to an electric field. The resistivity parameter is influenced by: texture and porosity, degree of cementation, temperature of the rock, clay content, body geometry, water content, temperature and salinity. Furthermore, under equal lithological conditions, there are some geological processes that cause an immediate variation of resistivity (Table 1). In general, many of these processes lead to a reduction of the resistivity as: clay alteration, dissolution, billing rock and saltwater intrusion.

Table 1. Resistivity rock values

Litotype	ρ (Ω m)	ϕ (%)
Water	10 - 100	
Sea water	2 - 3	
Sandstones	200 - 5000	7 - 30
Clays	1 - 50	40 - 70
Limestones	300 - 10000	2 - 30
Alluvial deposits	50 - 1000	15 - 60
Dolomite	500 - 10000	2 - 20
Sand and Gravel	70 - 700	30 - 60
Granites	1000 - 20000	0.2 - 0.8
Marl	100 - 500	8 - 15
Pyroclastics	50 - 600	15 -60
Igneous rocks	100 - 10000	30 - 10
Surface soil	10 - 200	60 - 90
Tuffs	150 - 900	10 - 40

The ERT 3D models, acquired in the time, allowed to obtain a 4D monitoring of saline intrusion in a specific subsurface "volume." The resistivity final model within that volume defines the electrical characteristics and the geometry of the subsurface, focusing the intrusion of salt waters and its relationships with the fresh waters. The results of this study lead to the reconstruction of a three-dimensional model of the water bodies in coastal plains in order to understand the extent of the phenomenon of saltwater intrusion in time and space. So, we can also evaluate the ratio volumetric between saltwater and freshwater. The following section describes the results of a 3D ERT experiment obtained in the coastal alluvial plain of the Volturno River (Figure 8) to assess changes in the freshwater-brine interface. One main aim was to investigate spatial and temporal variations of groundwater salinity. Data acquisition collected in May -October 2013 and in May - October 2014.

The instrumentation used for the measurement of the resistivity consists of two parts: one for the measurement of the current intensity (I) injected into the ground through the electrodes A and B and one for the measurement of the potential difference (ΔV) between the electrodes MN.

The electrodes are connected through multichannel cables, adopting the Wenner-Schlumberger reciprocal array configuration. This type of arrangement is hybrid between the Wenner and Schlumberger arrays [19]: during the acquisition, the wiring is continuously changed, so that the spacing between the 'potential electrodes' remains constant, while the distance between the 'current electrodes' increases as a multiple n of a. The value of n, in this case, is given by the ratio between the distance of the electrodes A-M (or N-B) and the spacing between the electrodes of potential M-N. For this array, the distribution of the measurements is comparable with the Wenner array, but the horizontal fold is better. The choice of such arrangement provides the maximum resolution respect to electrodes interval.

The selected area, geometrically similar to a rectangle of about 4600 square meters (length 115m and width 40 m) was surveyed for 4D monitoring of the salt wedge. The geoelectric survey was performed by acquiring 9 profiles. These profiles were arranged one by one parallels with a line spacing of 5 m and using a total of 216 electrodes.

The resistivity data processing has permitted to define the electrical characteristics and geometry of the subsurface allowing to spatially delimit the intrusion of salt water and the relationships with the brackish water. 3D models obtained by the inversion also showed seasonal variation (in time) of these relationships. The analysis of four models has shown very small variations of resistivity a) 4,36-31,2 Ωm; b) 4,66 – 30,36 Ωm; c) 7,45 – 32,7 Ωm d) 3 – 64,7 Ωm. The low resistivity zone (Figure 9), located in the lower and side part of the models, has been interpreted as salt water related to marine intrusion. The middle sector of the models presents resistivity values compatible with that ones typical of the fresh aquifer. In detail, changes can be observed between the volumes of brackish water and fresh-water during the spring and autumn surveyed seasons.

In particular, the results highlight lower resistivity values in autumn (October 2013-2014) in the shallow sector, ascribed to an increase of salt intrusion. This phenomenon is attributed to a lower contribution of fresh water in the aquifer due to the reduction of rainfalls during the summer season. The comparison, instead, with the two resistivity models acquired in May 2013 and 2014 showed no significant changes. The results of this study have led to a reconstruction of a three-dimensional model of the water bodies in coastal plains in order to understand the extent of the phenomenon of saltwater intrusion in time and space, resulting in qualitative and quantitative analysis of the volume of used water.

These analysis show the shallow fresh-water should not be utilized because the pumping could increase intrusion of the sea water and therefore this would produce damage at the native species plant. Finally, a monitoring program, supported by a careful management of resources, is necessary to prevent the worsening of the sea water intrusion avoid strongly hydrogeological disequilibrium.

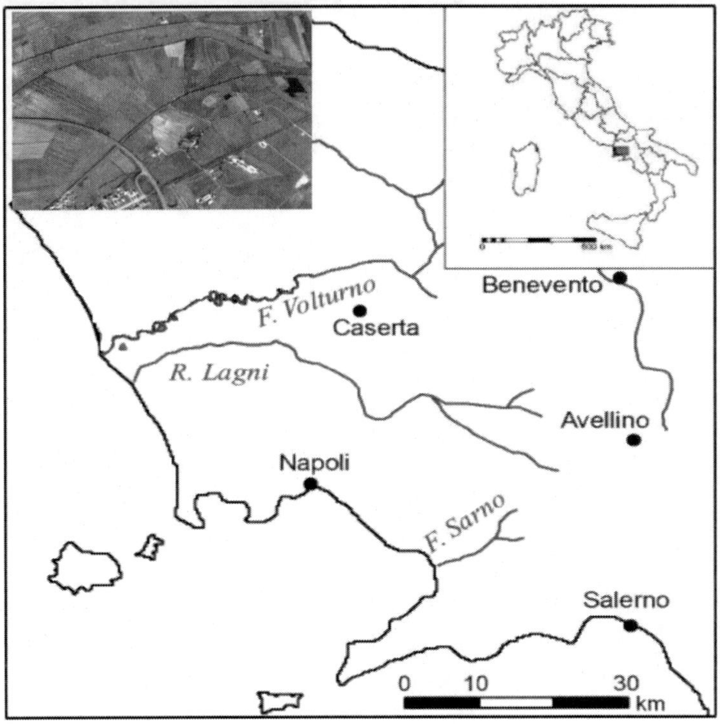

Figure 8. Location of the study area. Red lines identify the 3D ERT grid.

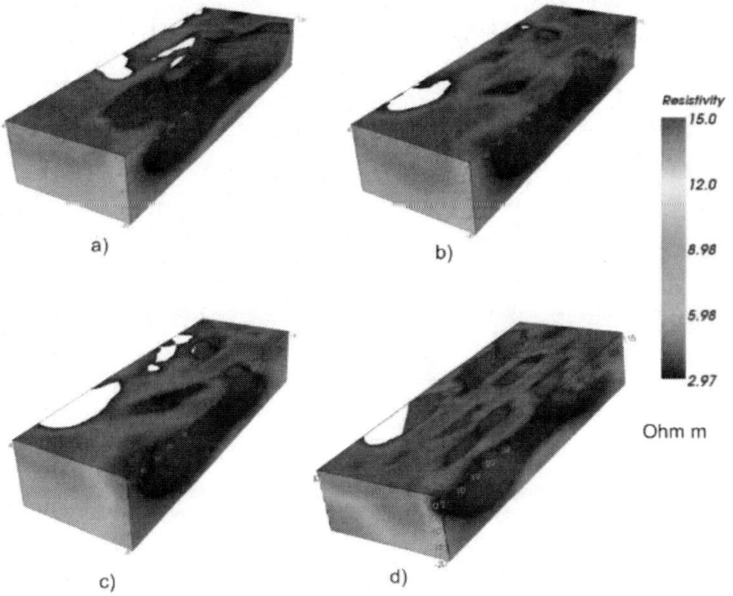

Figure 9. 3D resistivity contour plot referred to salt water (resistivity 4-15 Ωm): a) May 2013; b) October 2013; c) May 2014 d); October 2014.

REFERENCES

[1] European Commission, (2002). Integrated Coastal Zone Management (ICZM).

[2] Di Fiore, V., Cavuoto, G., Punzo, M., Tarallo, D., Pelosi, N., Giordano, L., Alberico, I., Marsella, E., Mazzola, S., (2015). Joining Up Land and Marine Seismic data: Case study from Procida and Ventotene Islands (Tyrrhenian Sea, Italy). *Geophysics*, 80(6), 153-166.

[3] Paffenholz, J., (1995). Method for deriving water bottom reflectivity in dual sensor seismic surveys. WO1995008782 A1. Western Atlas International, Inc.

[4] Barlow, P. M., (2003). Groundwater in fresh water-salt water environments along the Atlantic seaboard. *Circular* 1262 USGS.

[5] Chitea, F., Georgescu, P., Ioane, D., (2011). Geophysical detection of marine intrusion in black sea coastal areas (Romania) using VES and ERT data. Geo-Eco-Marina 17, 95-102.

[6] Al-Sayed, E. A., El-Quady, G., (2007). Evaluation of sea water intrusion using the electrical resistivity and transient electromagnetic survey: case study at Fan of Wadi Feiran, Sinai, Egypt. EGM 2007 International Workshop, Italy.

[7] Binley, A., Henry Poulte, S., Shaw, B., (1996). Examination of solute transport in an undisturbed soil column using electrical resistance tomography. *Water Resources Research* 32, 763-769.

[8] Binley, A., Shaw, B., Henry Poulte, S., (1996). Flow pathways in porous media: elettrical resistance tomography and dye staining image verification. *Measurement Science Technology* 7, 384-390.

[9] Daily, W., Ramirez, A., LaBrecque, D., Nitao, J., (1992). Electrical resistivity tomography of vadose water movement. *Water Resources Research* 28, 1429-1442.

[10] Schima, S., LaBrecque, D. J. and Miletto, M., (1993). Tracking fluid flow in the unsaturated zone using cross-borehole resistivity and IP. *Proceedings of the Symposium on the Application of Geophysics to Engineering and Enviromental Problems* 1993, San Diego, CA, 527-543.

[11] Daily, W., Ramirez, A., LaBrecque, D. and Barbe, W., (1995). Electrical resistance tomography experiments at the Oregon Graduate Istitute. *Journal of Applied Geophysics* 33, 227-237.

[12] Daily, W. and Ramirez, A., (1995). Electrical resistance tomography during in-situ trichloroethylene remediation at the Savannah River site. *Journal of Applied Geophysics* 33, 239-249.

[13] LaBrecque, D. J., Ramirez, A., Binley, A. and Schima, S. A., (1996). ERT monitoring of environmental remediation processes. Measurement Science Technology 7, 375-383.

[14] Stucchi, E., Zgur, F. Baradello, L., (2005). Seismic land-marine acquisition survey on the Great Ancona Landslide. *Near Surface Geophysics* 3, 235-243.

[15] Kaufmann, R. D., Xia, J., Benson, R. C., Yuhr, L. B., Casto, D. W., Park, C. B., (2005). Evaluation of MASW Data Acquired with a Hydrophone Streamer in a Shallow Marine Environment. *Journal of Environmental and Engineering Geophysics* 10, 87–98.

[16] Vicinanza, D., Guida, A., Ferrante, V., Ciavola, P., (2010). Performance of a beach dewatering system-Chiaiolella Beach, Procida Island, Italy. *Journal of Coastal Research* 26(4), 753–761.

[17] Communal Administration of the city of Ventotene, (2013). Indagini geologiche della Spiaggia "Cala Nave." Technical office, Municipality of Ventotene/.

[18] Fedele, L., Morra, V., Perrotta, A., Scarpati, C., Sbrana, A., Putignano, M.L., Orrù, P., Schiattarella, M., Aiello, G., Budillon, F., Conforti, A., D'Argenio, B., (2012). Note illustrative della Carta Geologica Regionale alla scala 1:10.000. Isole di Procida e Vivara. Regione Campania, Napoli.

[19] Pazdirek, O., Blaha, V., (1996). Examples of resistivity imaging using ME-100 resistivity field acquisition system. In: EAGE 58th Conference and Technical Exhibition Extended Abstracts, Amsterdam, P050.

In: Geophysics
Editor: Gemma Aiello

ISBN: 978-1-63484-831-2

Chapter 3

Marine Geophysics of the Naples Bay (Southern Tyrrhenian Sea, Italy): Principles, Applications and Emerging Technologies

Gemma Aiello [*] ***and Ennio Marsella***
Istituto per l'Ambiente Marino Costiero (IAMC),
Consiglio Nazionale delle Ricerche (CNR), Napoli, Italy

Abstract

Marine geophysics of the Naples Bay is discussed focusing particularly on some principles, applications and emerging technologies in seismic and magnetic marine data. The case histories of Somma-Vesuvius volcanic complex, Phlegrean Fields offshore and Ischia and Procida islands offshore (Naples Bay) are here discussed. Seismo-stratigraphic techniques and methodologies are explained, with the example of the Naples area, where the widespread volcanic activity during Quaternary times has disallowed for the application of a classical stratigraphic approach due to the occurrence of interlayered sedimentary sequences and interstratified volcanic bodies. Principles, applications and emerging technologies of marine geophysical data in the Naples Bay are considered with respect to the data acquisition, processing and interpretation, taking into account some historical aspects towards new technologies. Among the sources tested in the Naples Bay there are the explosives the Sparker and the Watergun while the details to study geomorphological data have been analyzed through the Surfboom and the Sidescan Sonar. MEAS (Multispot Extended Array Sparker) seismic source has been largely used to acquire large database of single channel reflection seismics. More recently, by means of Multitip SAM96, SAM400 transducer high resolution seismic data in the Bay of Naples have been recorded and interpreted in detail. Significant correlations among seismic and magnetic data have been attempted based on comparative analysis of seismic and magnetometric datasets recorded in the Naples Bay.

[*] Corresponding author: Gemma Aiello. Istituto per l'Ambiente Marino Costiero (IAMC), Consiglio Nazionale delle Ricerche (CNR), Calata Porta di Massa, Porto di Napoli, 80133, Napoli.

1. INTRODUCTION

The seismic stratigraphy of the Naples Bay is here resumed and discussed from the old to the new data. We focus on case histories located in Somma-Vesuvius offshore, Gulfs of Naples and Pozzuoli (Phlegrean Fields offshore) and Neapolitan islands (Ischia and Procida). Emerging technologies in marine data acquisition, processing and interpretation of the Naples Bay (southern Italy) are shown in this chapter, dealing in particular with the contribution of seismo-stratigraphic and marine magnetic data, whose theoretical aspects will also be discussed. Seismic stratigraphy and marine magnetics of aforementioned selected case histories are here shown, representing excellent applications of geophysical methodologies.

Seismic exploration is commonly performed by means of sources than can generate elastic waves from a rapid expansion of underwater gas bubbles. These bubbles can control many pulses that take the form of double exponential spikes of gradually decreasing amplitude [44].

Several technologies can be used in order to produce an acoustic pressure wave into water, such as free-falling weights chemical explosives piezoelectric or magneto-resistive sources sparkers boomers airguns and waterguns. Each source has a precise signature and wave frequency that can be considered optimal as a function of depth and resolution. The main characteristics of a seismic source is to produce a single high-energy spike that is detectable, despite the presence of a noise, after crossing the portion of the seabed that we wish to study. A broad range of frequencies can be reproduced, as well as a broad range of waveforms can be generated as a function of frequency-dependent absorption of elastic waves and nearby boundaries presence.

Seismic sources for offshore investigation may be impulsive, providing a short-lived burst of elastic wave energy and swept-frequency, which triggers a low-amplitude sinusoidal signal. Impulsive sources such as explosives can cause damages to marine flora and fauna; for this reason towed sources activated for only few seconds must be preferred. The type of source could be chosen depending on the required resolution and signal penetration. Vibration of piezoelectric and magnetic materials electric pulses or pressured fluid discharge often organized into arrays can be considered as good seismic sources whose signature spectra and energy output are variable.

Sparkers [90] and Boomers [75] systems are respectively based on an electrode array powered by high voltage capacitor bank and by an electromagnetic source. Sparkers and boomers can generate seismic energy to explore continental margins when there are near surface or deep-towed seismic sources (10-50 m off the sea). Moreover, Boomers with pulse length of 0.1-0.2 msec can be used to investigate very shallow waters. Sparker systems can produce low-frequency acoustic waves (the maximum frequency contained in the spectrum of acoustic signal is approximately 2000 Hz) that can penetrate several hundred of meters of sediments.

One of the most interesting seismic sources is the Multispot Extended Array Sparker (M.E.A.S.) that consists of sparker electrodes disposed on a square metal cage. This kind of system, patented by Institute of Oceanology of Istituto Universitario Navale of Naples (Italy) allows obtaining a good signal penetration and high resolution seismic data with relative small energy use. The M.E.A.S. signal is a short impulse with a large frequency spectrum content (Figure 1).

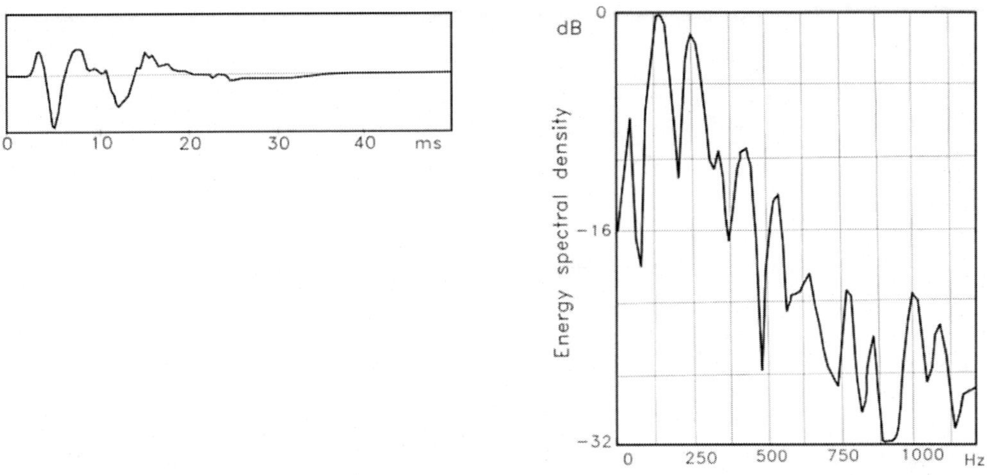

Figure 1. Signature and spectrum of Multispot Extended Array System (modified after Mirabile et al., 1991).

Mirabile et al. (1991) [102] tested the acquisition geometry in order to reduce a superimposing of source signal with return echoes that respect the "far-field" condition. Moreover, these authors have demonstrated the utility of some techniques for signal deconvolution in order to produce the so-called seismic profiles "deghosting." Seismic reflection data require complex series of numerical treatments to increase the signal/noise ratio of a single profile as well as obtaining a high resolution seismic section to improve the geological interpretation.

A more recent technology is the Sparker source SAM that is characterized by a varying number of electrodes that can be disposed as "dual-in-line" (SAM96) and "planar array" multi electrode electro-acoustic source (SAM400/800; Figure 2).

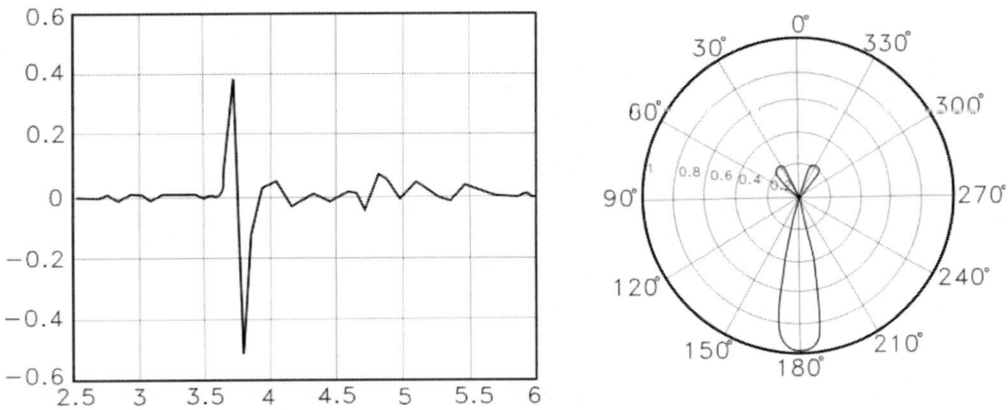

Figure 2. Left: Signal (received by sub-surface hydrophone) generated by SAM system at firing energy 200 J duration of primary impulse 0.3 ms (modified after Corradi et al., 2009); Right: Square radiation diagram that shows the high system directivity.

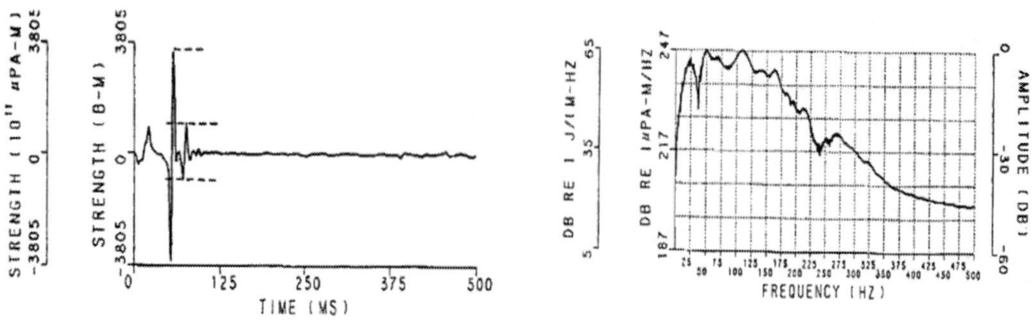

Figure 3. Example of water-gun signature in far field and the frequency spectrum (modified after Ranieri and Mirabile, 1991).

Other seismic sources are the Airgun and the Watergun, recently used in the Naples Bay in submarine geological mapping and basin studies [51, 6, 8, 9]. The Airgun produces high energy short seismic pulses by means of a discharge of compressed air into water, related with a wide range of pulse shapes and source spectra (Figure 3). On the other side, the Watergun produces the sudden collapse of a cavitation volume into water that is proportional to kinetic energy of the water plug.

Since the '70s until now many attempts have been carried out in order to improve seismic technologies performance, data acquisition and processing. In the practice of seismic prospecting, Sparker systems technologies were widely analyzed using different acquisition systems. Some ones consist of a single electrode hotter than a mass electrode, other ones of more electrodes over distributed mass (eg. Sparker Teledyne and Sparker EG&G).

De Vita et al. (1979) [64] tried to identify based on experimental data which configuration is more appropriate (single electrode or multi-electrode) using the fundamental equations for the design of an "array." Sparker signals are the base band signals, transitory and continuous spectrum. Based on these measurements it has been demonstrated that energy should never exceed 400 joules/electrode to achieve the best compromise between resolution and electro-acoustic performance.

Ranieri and Mirabile (1991) [131] have reported technical and scientific results obtained through the geophysical survey of the deep geological structure of the Phlegrean Fields volcanic complex. It was aimed at improving the knowledge on technologies and sources that are more appropriate for the investigation of the continental margins, particularly in complex volcanic areas like the Gulf of Naples [82].

Among the sources tested in studies of the Gulf of Naples there are the explosives [101], the Sparker and the Watergun, while the details to study geomorphological data were analyzed through the Surfboom and the Sidescan Sonar. MEAS (MultiSpot Extended Array Sparker) [102] seismic source (12 and 16 kJ) consists of an array of 36 (6x6) electrodes placed inside a metal cage in a square size 4.5x4.5 m, spaced 0.75 m and fed in phase.

The energy used by the MEAS has a pulse of short duration in the order of 10 milliseconds and a significant spectral content up to 1000 Hz with maximum energy output around 150-200 Hz. Each echo corresponds to an acoustic discontinuity (impedance contrast) that can generally be interpreted in geological terms.

MEAS system has been largely used in order to acquire a large database of single channel reflection seismics in the Bay of Naples [100, 91,102]. Recently, by means of Multi-tip

SAM96 (0.1-1 kJ), SAM400 (1-4 kJ) transducer it was possible to record high resolution seismic data in the Bay of Naples both in coastal and deep sea research [45]. Some evidences on magnetic field anomalies in the Gulf of Pozzuoli come from the magnetic map of Galdi et al. (1988) [83] reporting a NE-SW interruption of main regional trend where some circular local anomalies are related to products of post-calderic volcanic activity [134].

The integrated marine magnetics of the Naples Bay is resumed and discussed from the old to the new data. We will focus on some case histories of magnetic data recorded in the Naples and Pozzuoli Gulfs (Southern Italy). Significant correlations between geophysical data have been carried out by using the comparative analysis of seismic and magnetometric datasets. A magnetometer usually measures the strength or direction of the Earth's magnetic field. These parameters can vary both temporally and spatially for various reasons, including discontinuities between rocks and interaction among charged particles from the sun and the magnetosphere. The most of technological advances dedicated to measure the Earth's magnetic field have taken place during World War II. The most common instruments are the fluxgate magnetometer the proton precession magnetometer the Zeeman effect magnetometer the the sensor-suspended magnetometer and the satellite magnetometer. The fluxgate and the proton precession magnetometers are effectively the most used for marine surveys since they are both cable drown. The fluxgate magnetometer was the first ship-towed instrument and it can measure vector components of the magnetic field. Its sensor consists of two magnetic alloy cores that are mounted in parallel configuration with the windings in opposition. The proton precession magnetometer consists of a sensor containing a liquid rich in protons surrounded by a coil conductor. The sensor is towed from the vessel through an armoured coaxial cable whose length depends on vessel length and seafloor depth. Circulating current within the coil generates a magnetic field of approximately two orders of magnitude the Earth's field. In this way, one proton each ten will follow the coil positioning. Stopping the induced magnetic field, the protons will align according to the Earth's magnetic field through a movement of precession.

The proton precession magnetometer is one of the most used for offshore surveys and it records the strength of the total field by determining the precessional frequency (f) of protons spinning about the total field vector as it follows:

$$F = \gamma_p F/2\pi \qquad \text{(equation 1)}$$

where γ_p is the gyromagnetic ratio of the proton uncorrected for the diamagnetic effect, so that knowing it from laboratory measurements, the total field in nanotesla can be calculated as:

$$F = 23.4866xf \qquad \text{(equation 2)}$$

The Total Magnetic Field calculated through the equation (2) is stored by magnetometer into a string of data containing position data that are displayed as a x,y chart. The signal frequency is measured on a time span of 0.5 seconds when the signal-noise ratio is highest. It is necessary to use two orthogonal coils in order to ensure a maximum initial value of proton precession. The measured field must be corrected with respect to the regional magnetic field in order to evaluate the magnetic anomalies.

The proton precession magnetometer was largely used to explore magnetic anomalies in the Bay of Naples. Interesting examples of magnetic data acquisitions in Naples and Pozzuoli bays are reported in Galdi et al. (1988) [83]; Secomandi et al. (2003) [141]; Aiello et al. (2004) [5] and will be discussed in the following paragraphs. As shown by the maps of Galdi et al. (1988) [83] both positive and negative anomalies have been detected using a Geometrics G-856 magnetometer model. Globally the area shows an interruption of the regional trend from NE-SW where circular anomalies are probably connected to a post-calderic activity of the Phlegrean Fields. Moreover, the internal area of the Pozzuoli Bay is characterized by a negative anomaly that increases towards the south. Conversely, in the external area there is mainly an alternance of positive and negative anomalies with a dominance of positive values near the area of Bagnoli. For a detailed analysis of the magnetic anomaly field of the volcanic district of the Bay of Naples see Secomandi et al. (2003) [141]. Recently, Aiello et al. (2004) [5] presented a high resolution magnetic anomaly map of the Bay of Naples based on data acquired during the oceanographic cruise GMS2000-5 performed in October-November 2001 onboard of the R/V Urania using the EG&G Geometrics proton magnetometer G-811.

2. Geologic and Geophysical Setting

2.1. Somma-Vesuvius

In the Gulf of Naples one of the most important lineaments is the Somma-Vesuvius volcano. The Vesuvius volcano has been intensively studied with respect to the eruptive events the recent seismicity the geochemistry and the ground movements of the volcano and the related volcanic hazard [34, 136, 35, 77, 97, 135, 139, 145] The Somma edifice started to grow after the eruption of the Campanian Ignimbrite deposits. Its eruptive activity ranges from “Pomici di Base” (18 ky) and the “Pomici Verdoline” plinian eruptions, enabling the collapse of the Somma edifice and the consequent calderization. Three main plinian eruptions (“Mercato,” “Avellino” and “Pompei”) followed among 7900 B.C. and 79 A.D. The total magnetic field map of the Somma-Vesuvius volcano (Figure 4) has shown interpretative elements that have given an indicated value for the trend of volcanites in the volcanic complex’s peripheral areas [34]. A main sub-circular anomaly is centred on the volcano; two positive appendages diverge towards SE and SW. They might correspond to a great thickness of lava products, possibly in pre-existing depressions of the sedimentary basement of the graben of the Campania Plain. These products should have enhanced the occurrence of an elongated magnetized body, which tends to move towards the Naples Bay from the Vesuvius volcano towards Torre del Greco; an alternative explanation would be the presence of a strip of eruptive vents, settled on a system of NE-SW normal faults [21, 80]. A new aeromagnetic map of the Vesuvius area has been produced (Figure 5) [120]. It is dominated by a large dipolar anomaly related to the volcano, showing a southwards trending elliptical shape. A narrow anomaly is located on the western flank of the edifice and small anomalies are located on the south-eastern slope of the volcano. High frequency anomalies occur in the area surrounding the edifice related to the high cultural noise of this densely inhabitated area.

Emerging technologies in Somma-Vesuvius area include the use of tomographic techniques for the modeling of magma chambers of the volcano and the implementation of seismo-stratigraphic knowledge through seismic reflection studies mainly finalized to geologic and volcanic hazard prevision.

The high density of population of the Vesuvius hinterland makes this sector an area of high volcanic and geologic hazard which after 60 years from the last eruption has stimulated many volcanological and geophysical researches aimed at a better comprehension of the volcano structure. Only through a detailed knowledge of the deep structure it is possible to advance real hypotheses on the possible eruptive dynamics which are basic in a territory zonation finalized to a reduction of the volcanic hazard.

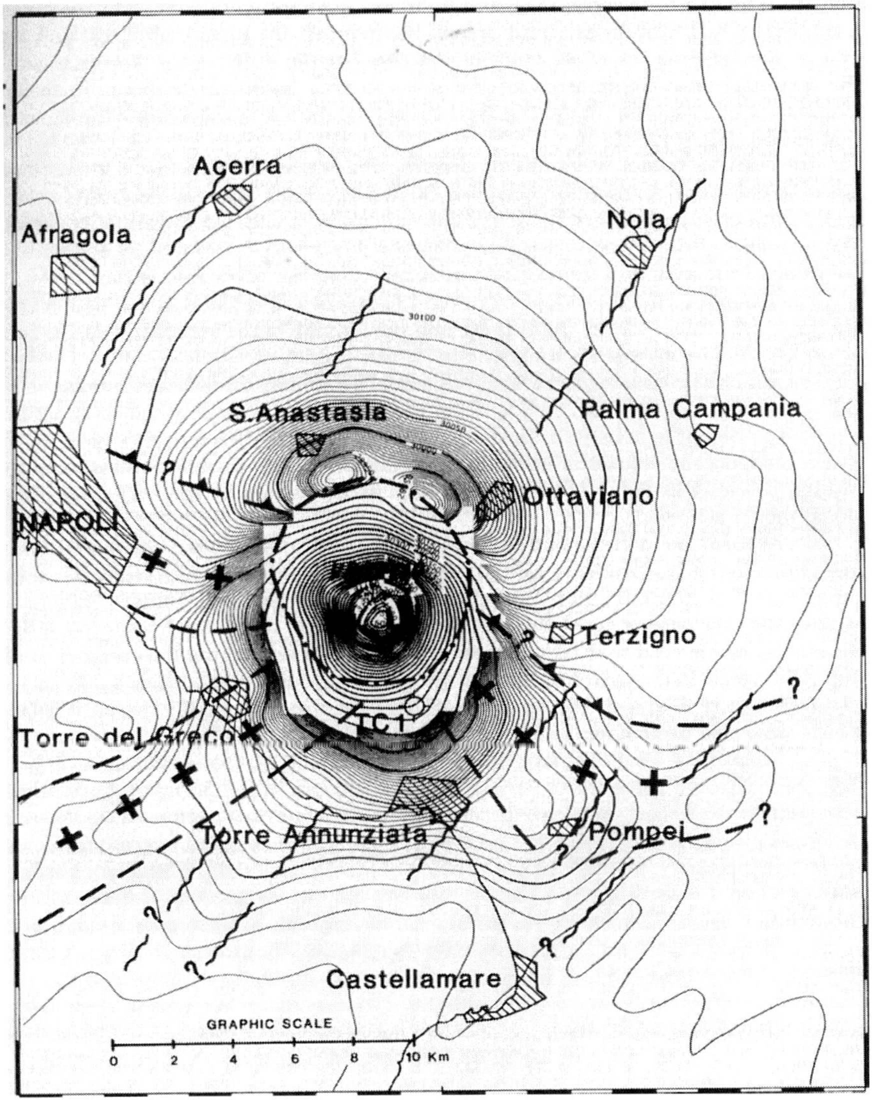

Figure 4. Total magnetic field map of the Somma-Vesuvius volcano with sketch structural interpretation (modified after Cassano and La Torre, 1987).

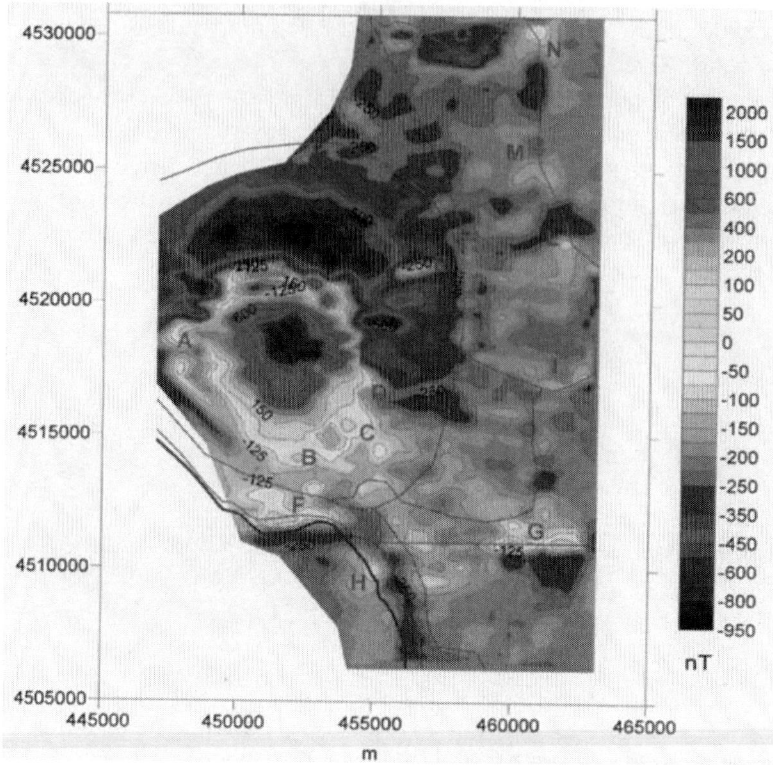

Figure 5. Aeromagnetic map of the Vesuvian area (modified after Paoletti et al., 2005).

Nowadays the Vesuvian area is divided into three zones (red, yellow, blue) based on the degree of volcanic risk, decreasing with the increasing of the distance from the volcanic edifice. The boundaries among the different zones derive from a probabilistic analysis of the eruptive history of the volcano by assuming as a realistic scenario a middle-degree eruption as the eruption of 1631. The lacking of an eruptive model constrained by a detailed structural model of the volcanic complex has often determined a notable difficulty in the choice of classifications which are becoming a compromise between administration limits and volcanic risk.

For these reasons the geological models on the Somma-Vesuvius structure evolved during time becoming more and more detailed. The studies were mainly aimed at the definition of the depth and dimension of the magmatic tank. The knowledge on the shallow structure of the volcano has been improved after the Agip drilling due to geothermal findings in the Trecase territory concluded on March 1981. The Trecase well was drilled up to a depth of 2 km from the field plan, showing the whole volcaniclastic sequence and reaching the top of the Mesozoic carbonate basement (Figure 6). In the 1987 Cassano and La Torre [34] have interpreted the Bouguer anomaly map of the Campania Plain with a two-dimensional model along the Acerra graben-Rovigliano inlet profile, determining the depth of the top of the carbonate horizon. In the 1996 the projects Tomoves, Mareves and Broadves started which have furnished tomographic images of the longitudinal waves of the volcanic complex up to a depth of 4 kilometers [149, 150, 93, 60]. Tomographic studies have individuated a high velocity zone below the volcanic complex interpreted as a structure of dykes. The data recorded during these projects have suggested two possible locations of a conversion horizon, supposed to be the top of a magmatic chamber [149, 17, 59]. In the 1999 Bruno and Rapolla [28] have re-processed deep seismic data recorded by Agip along the flanks of Somma-Vesuvius highlighting a deep fault system in the south-eastern sector of the volcanic complex which controlled the volcano dynamics. Cubellis et al. (2001) [47] have presented a gravimetric model of the volcanic complex suggesting the occurrence of a solid magmatic body crossing the carbonatic basement. De Natale et al. (2005) [61] have purposed several models of shear waves of the volcanic complex up to a depth of 30 kilometers for a set of seismic stations located in the Vesuvius. These data have been obtained through a linear inversion of the average dispersal curves of the group and phase velocities from 0.2 sec to 30 sec. The purposed models have shown two velocity inversions at about 7-11 kilometers and below 11-20 kilometers, attributed to the occurrence of partially melted material. Nunziata et al. (2006) [112] have constructed a sketch model of the shear waves for the Vesuvius, starting from the Tyrrhenian sea and since to a depth of 135 kilometers. This model shows two partially melted zones respectively located at 8-9 kilometers and at 20-25 kilometers and a Moho depth at 15-17 kilometers.

Geochemical studies on the fluid and melted inclusions and on the isotopic rates [19, 20, 94, 92, 121, 126] have evidenced a feeding system of the volcanic complex at about 4-5 kilometers of depth, originating the Plinian activity of the Vesuvius volcanic complex.

The Somma-Vesuvius volcanic complex is located in the southern sector of the Campania Plain, a structural depression Plio-Quaternary in age between the eastern side of the Tyrrhenian sea and the Southern Apennines. This depression is controlled by NW-SE and NE-SW normal fault systems and is filled by sedimentary deposits ranging in age from the Early Pleistocene to the Holocene. During the Quaternary the south-western sector of the area was involved by a strong volcanic activity, as testified by the individuation of the volcanic districts of Phlegrean Fields, Ischia and Procida, Roccamonfina and Somma-Vesuvius.

The Somma-Vesuvius is a complex volcano, whose activity began about 400 ky B.P. [27]. The volcanic edifice consists of an oldest vent repeateadly collapsed (Somma) and a younger vent (Vesuvius). The two vents are linked by the Gigante valley, a narrow half-circled depression whose bottom is overlain by lava fluxes of different eruptions. The Somma is composed of lava flows and minor deposits of pyroclastic flows and fluxes. The Vesuvius cone is constituted of pyroclastic deposits of plinian and sub-plinian eruptions. The Somma caldera is bounded by normal faults originated during explosive eruptions. It has an elliptical shape whose main axis is E-W oriented. The caldera rims are well developed in the northern sector of the edifice as sub-vertical walls having a maximum height of 280 m. In the western and southern sectors the caldera rims are overlain by products younger than the event of 1631 D.C. [129]. The Vesuvius is a typical volcanic cone having an upper diameter of 450 m and a depth of 330 m [47]. The crateric rim shows a notable asymmetry being steeper in the north-eastern sector. The volcanic activity of the Somma-Vesuvius is of a mixed type characterized by alternating effusive and plinian eruptions. During the last 20 ky B.P. seven plinian eruptions happened at intervals of several thousands of years. The youngest one is the important eruption of 79 D.C., which destroyed the Pompei and Ercolano towns. It has been calculated that each plinian effect was capable to produce from 5 to 11 km^2 of pyroclastic materials.

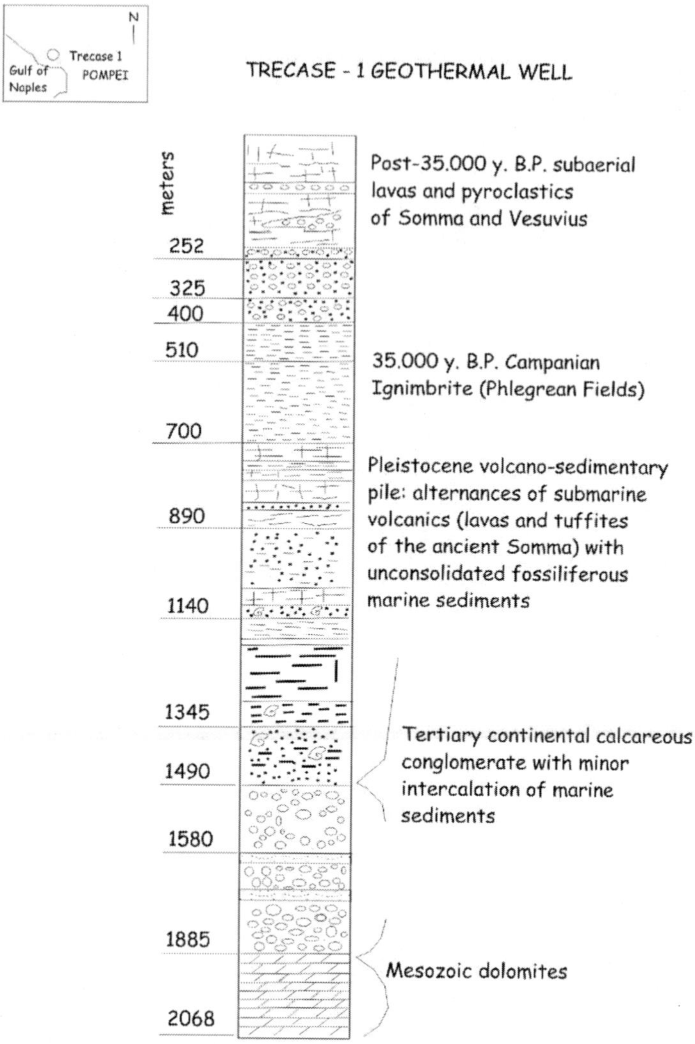

Figure 6. Stratigraphy of the Trecase 1 exploration well (modified after Brocchini et al., 2001). The well has drilled several volcanic complexes, respectively overlying and underlying the Campanian Ignimbrite pyroclastic flow deposits, representing a marker horizon dated at about 37 ky B.P.

The eruptive history of the Somma-Vesuvius complex may be distinguished in three periods [16]. The oldest one is also the longest one and pre-dated the Pompei eruption: historical data are lacking and the reconstructions have been carried out only based on geology. The second period covers a time interval ranging from 79 D.C. to 1631 D.C.: the available historical data are often incomplete

2.2. Phlegrean Fields and Gulf of Pozzuoli

The Phlegrean Fields are a volcanic district surrounding the western part of the Gulf of Naples, where volcanism has been active since at least 50 kyr [134], corresponding to a resurgent caldera having a diameter of 12 km and erupting the Campanian Ignimbrite (37 ky B.P.) and the Neapolitan Yellow Tuff (15 ky B.P.) deposits. A volcano-tectonic uplift of the caldera center has been suggested by coastal sediments ranging in age from 10 to 5.3 ky B.P., cropping out in "La Starza" marine terrace (Gulf of Pozzuoli) [74, 108].

The Quaternary volcanic area of the Phlegrean Fields is located in a central position within the graben of the Campania plain [50, 24]. The pre-calderic volcanic activity developed in correspondence with small and scattered volcanic centers, erupting trachytic pyroclastites and lavas. The corresponding outcrop located at Monte di Procida coastal cliff exhibits a thick subaerial pyroclastic sequence with several interbedded paleosols. It includes four main pumice fall beds, a welded pyroclastic flow deposit and dark lapilli and ashes erupted by the Fiumicello volcanic center (Procida island). The post-calderic volcanic activity developed during four main phases individuated through radiometric age determinations and palaeogeographic evolution in a time interval spanning from 35 ky B.P. to historical times, as established by deep wells located in Mofete and San Vito geothermal areas [134].

Gravimetric and magnetometric information available for the Phlegrean Fields has been summarized [34]. From N to S the most important gravimetric elements are the positive anomaly related to the carbonatic horst of the Massico Mount the negative anomaly of the Volturno graben and the positive gravimetric anomalies of Villa Literno and Parete. The total magnetic field map (Figure 7) [34] has evidenced a strong positive anomaly in the area of Monte di Procida genetically related to weaker anomalies in the Procida Channel and in the Procida and Ischia islands. This anomaly has been controlled by the occurrence of considerable volumes of lavas erupted from trachybasaltic and latitic eruptive centres of Procida. Another large magnetic anomaly characterizes the Astroni-Agnano volcanic area, probably as the result of overlapping lava bodies. A new aeromagnetic map supplements the northern sector of the Phlegrean Fields [119], allowing for a better geologic interpretation of the structural patterns and morpho-structural features in the Volturno Plain and in the Gulf of Pozzuoli. Main magneto-structural features are the caldera rims of the Neapolitan Yellow Tuff and the Torregaveta anomaly. A complex pattern of magnetic anomalies coincides with the Parete volcanic complex [8], while another isolated anomaly corresponds to the Volturno River.

Seismo stratigraphic and marine magnetic data of the Gulf of Pozzuoli are here presented based on the geologic interpretation of recently collected Sparker profiles [10, 11]. The geology and the volcanology of the Phlegrean Fields volcanic complex and its adjacent offshore (Gulf of Pozzuoli) are here improved through a new stratigraphic setting delineated through seismic stratigraphy. Offshore seismic units have not been yet discussed in detail in an active volcanic area such as the Gulf of Pozzuoli, where active volcanism is suggested by bradyseismic movements and small earthquakes. According to many geological interpretations this area represents the submerged border of the Phlegrean Fields caldera located in correspondence to the Nisida, Miseno and Pentapalummo banks [124, 98, 6]. Deep onshore drilling of the Phlegrean Fields volcanic units has been recently carried out at the Bagnoli quarter (Naples town) in order to clarify the nature of the deep volcanic units underlying the caldera [62].

Seismic image and rock properties of the Phlegrean Fields caldera have been recently discussed through joint analysis and tomographic images based on the Serapis experimental dataset in the Gulf of Pozzuoli [58]. Obtained results constrain the identification and the characterization of the volcaniclastic layer filling the shallowest part of the Phlegrean caldera.

Tomographic images have revealed a correlation among P-velocity anomaly at 0.5-0.75 km of depth splitted into two parallel arcs and the location of volcanic dykes buried volcanic structures and offshore fumaroles detected in the Gulf of Pozzuoli [54, 124, 99, 10, 11]. The analysis of the vertical component CMP gathers located in the Pozzuoli southern rim has revealed a strong amplitude reflection event at 0.6 – 0.7 sec (twt). It should be genetically related with a a thick trachitic and latitic lava layer at about 750 m of depth under tuff and tuffites post-caldera determined in the Mofete well (Phlegrean Fields) [58]. While the volcanic activity gradually became subaerial in the northern part of the caldera, in the southern sector (i.e., the present-day Gulf of Pozzuoli), the conditions of marine sedimentation persisted since the caldera collapsed except for the Wurmian phase (18 ky B.P.), when the sea level reached -120 m and all the area was emerged above the sea level [124].

The Gulf of Pozzuoli is an inlet having limited dimensions bounded seawards by several submerged volcanic banks (Pentapalummo Bank; Nisida Bank; Miseno Bank) [6]. A morphological sketch map representing several offshore morphological units is shown in Figure 9 [10]. These units include the inner continental shelf the central basin the submerged volcanic banks and the outer continental shelf (Figure 9). The inner shelf is located in the northern sector and grades into a central basin deep 100 m through a 50 m deep shelf break and a steep slope. The Pozzuoli central basin is limited by a belt of submarine volcanic highs joining the outer shelf deep 140-160 m.

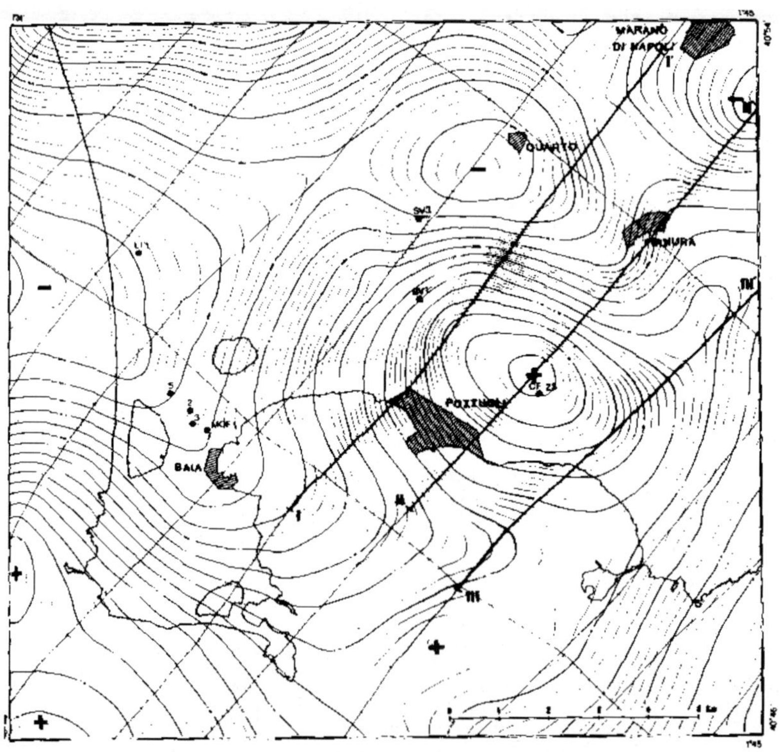

Figure 7. Total magnetic field map of the Phlegrean Fields area (modified after Cassano and La Torre, 1987).

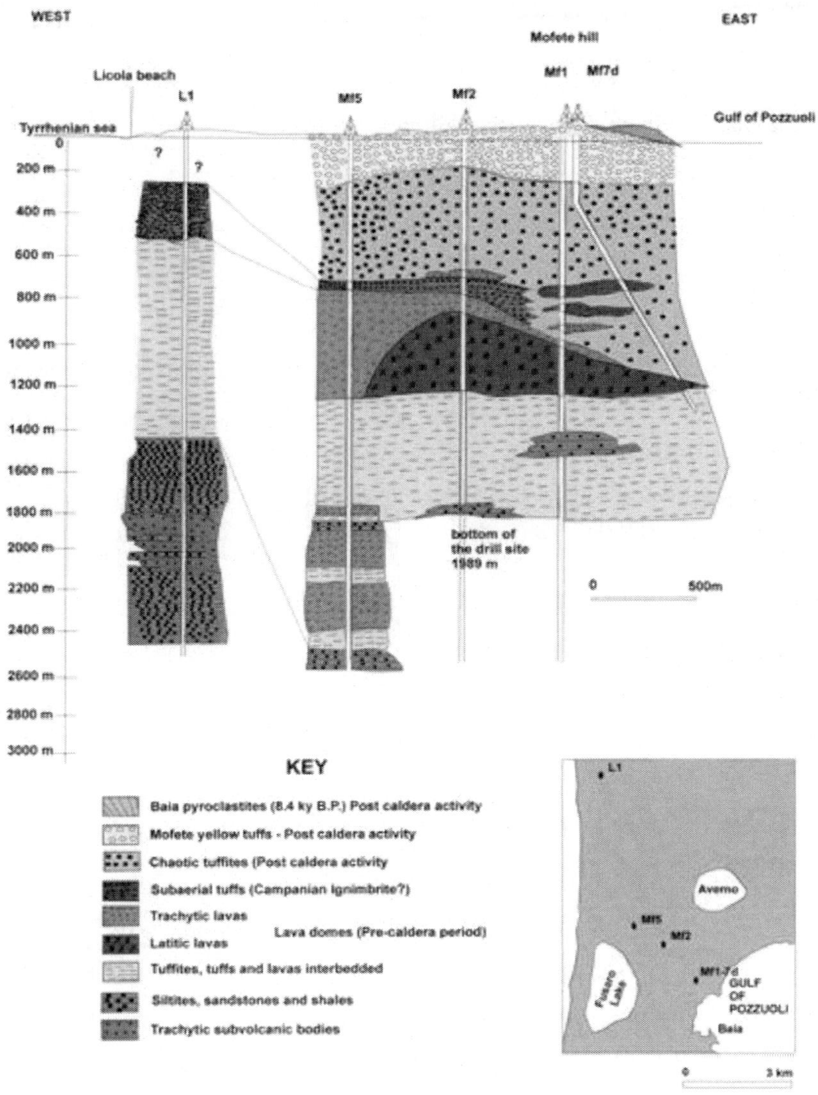

Figure 8. Sketch stratigraphic section of the Mofete geothermal area in the Phlegrean Fields (modified after Rosi and Sbrana, 1987).

2.3. Ischia and Procida

Located westwards of the Phlegrean Fields caldera, the Ischia island is formed by volcanic rocks erupted from different centres. The oldest outcrops date back to about 150 ky B.P., while the most recent eruptions occurred in 1302 A.D. The central sector of the island is characterized by the Monte Epomeo structure (787 a.s.l.), a structure uplifted by the resurgence of the caldera formed after the large explosive eruption of the Monte Epomeo Green Tuff [147, 12, 106]. The resurgence started about 33 ky B.P. producing the Mt Epomeo structure, whose edges are marked by NW-SE, NE-SW and E-W normal fault systems [115, 147, 32]. The total average uplift of about 800 m, inferred from the present height of marine deposits on Mt. Epomeo, occurred as a discontinuous process at an average velocity of about 3 cm/year. The resurgence interpreted as due to the pressure increase of a shallow magmatic body was accompanied by volcanic activity out of the resurgent block with dome, spatter cones and tuff ring emplacement. The geological evolution ended with the dismantling of the Ischia southern slope by volcanic avalanches [144]. Historical seismicity was confined in the northern sector of the island, while moderate volcano-tectonic processes occurred in the western slope of the island. The supposed trachytic intrusion whose top is located at 2 km of depth is responsible for the hydrothermal activity still active in some sectors of the island [32].

The Procida island, located in the Naples Bay, shows a geologic history genetically related with continental Phlegrean Fields and Ischia island. Procida and Vivara are two small volcanic islands, covering a whole surface of 5 km^2, located in an intermediate position between the active calderas of the Phlegrean Fields and Ischia island. Several stratigraphic and geologic studies have been carried out, indicating the occurrence of several pyroclastic units linked to the eruptive activity of the Ischian and Phlegrean complexes, interstratified in the volcanic products erupted by local centres [122, 69, 130, 123]. Six volcanic centres have been recognized including the Vivara volcano the Pozzo Vecchio volcano the Terra Murata volcano the Fiumicello volcano and the Solchiaro volcano. In the Ischia Channel relict volcanic morphologies ("La Catena" and "Formiche di Vivara"). The volcanic activity followed the tectonic phase of regional extension involving the Campanian area starting from the Pliocene and resulting in the formation of the graben of the Campania Plain [137, 73, 67]. A sketch map of the Phlegrean volcanic district (Phlegrean Fields, Ischia and Procida) has been reported, including the main tectonic structures, caldera rims, crateric rims and vent location.

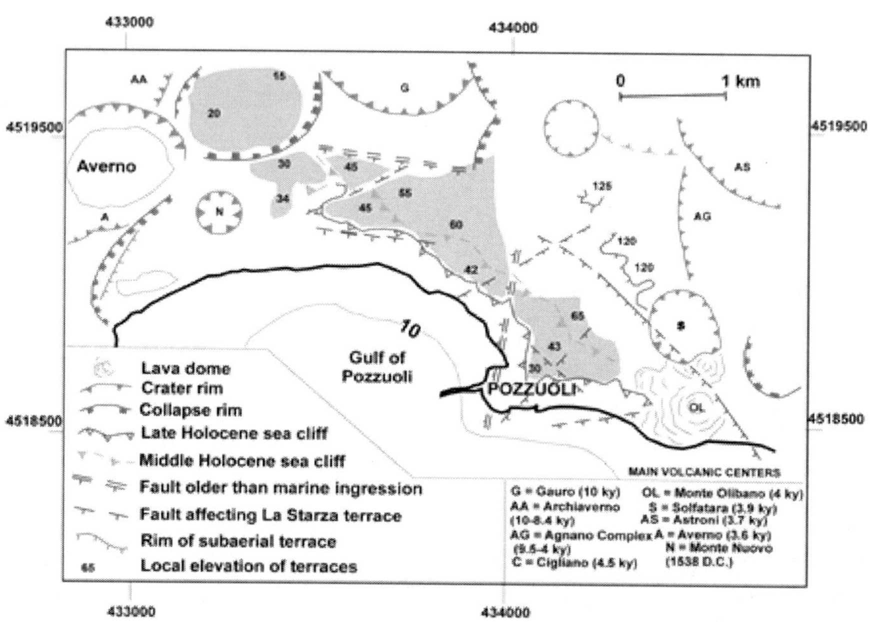

Figure 9. Sketch morphological map of the Gulf of Pozzuoli (modified after De Pippo et al., 1984).

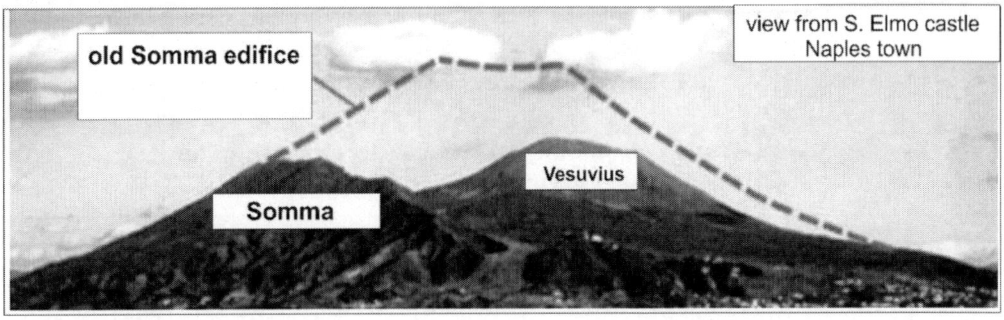

Figure 10. Qualitative reconstruction of the original profile of the old Somma volcano (modified after Cioni et al., 2009).

3. Remarks on the Volcanology of the Campania Volcanoes

3.1. Somma-Vesuvius

The Somma-Vesuvius is a stratovolcano of average dimensions, reaching a maximum height of 1281 m. It is composed of an oldest volcano (Somma) whose top has been downthrown generating a caldera and of a youngest volcano (Vesuvius volcano) later grown in the calderic depression. A qualitative reconstruction of the original profile of the old Somma volcano has been constructed (Figure 10) [40]. The Somma edifice is the relict of the northern flank of the old volcanic edifice.

The volcanic activity in the Somma-Vesuvius area started at least 400 ky B.P. based on the age dating of lavas found in deep exploration wells (Trecase 1 well) [27]. The eruptive history of the Somma-Vesuvius volcano can be resumed as it follows (http://ingv.it).

Volcanism Oldest than 19 ky B.P.

The great Phlegrean eruption of the Campanian Ignimbrite has provoked the burial of the most part of the Campania region under a thick coverage of tuffs. The Somma edifice started to grow on these deposits as a consequence of effusive eruptions, subordinately explosive, having a low energy. This activity lasting 19 ky B.P. has controlled the formation of the Somma volcanic edifice. The northern part of the old edifice is still well preserved and is represented by the present-day Monte Somma.

Volcanism Ranging in Age between 18.3 and 16 ky B.P.

This period of activity is dominated by two main Plinian eruptions: the eruption of the Pomici di Base (18.3 ky B.P.) and the eruption of the Pomici Verdoline (16 ky B.P.). Lavas erupted by subordinate effusive episodes are interstratified with the deposits of these two eruptions. The collapse of the Somma edifice was coeval with respect to the first plinian eruption of the Pomici di Base and was accompanied by the Vesuvius caldera formation.

Volcanism Ranging in Age between 8 ky B.P. and 79 D.C.

During this period of activity three Plinian eruptions verified: the eruption of the Pomici di Mercato (8 ky B.P.) the eruption of the Pomici di Avellino (3.8 ky B.P.) and the Pompei eruption (79 D.C.). The volcanic products of six subplinian eruptions ranging in age from the Avellino and the Pompei eruptions alternating with long periods of quiescence evidenced by paleosols or erosional/depositional hiatus have been recognized by volcanological studies.

Table 1. Sketch chronology of the eruptive events in the Somma-Vesuvius volcanic complex

Age dating (ky B.P.)	Type of volcanic activity	Somma-Vesuvius eruption
18.3	Mainly explosive	Pomici di Base
ranging between 18.3 and 16	Mainly effusive	Lavas spreading from subordinate lateral centres
16	Mainly explosive	Pomici Verdoline
8	Mainly explosive	Pomici di Mercato
ranging between 8 and 3.8	explosive/effusive	At least two subordinate eruptions
3.8	Mainly explosive	Pomici di Avellino
ranging between 3.8 and 79 D.C.	explosive/effusive	At least two subordinate eruptions
79 D.C.	Mainly explosive	Pompei eruption
472 D.C.	explosive/effusive	Pollena eruption
1631 D.C.	Mainly explosive	1631 eruption
1631-1944	Explosive/effusive	18 strombolian cycles
1944	Mainly explosive	Final eruption

Volcanism Ranging in Age between 79 and 1631 D.C.

The activity of this period included almost two subplinian eruptions: the Pollena eruption (472 D.C.) and the 1631 eruption. They are related with a set of small effusive and explosive eruptions having a low energy, originating lava flows along the western and southern flanks of the volcano and levels of strombolian scorias of medieval age.

Volcanism Younger than 1631 D.C.

After the eruption of the 1631 and since the 1944 the Vesuvius was characterized by an activity having an open conduit. During this period 18 strombolian cycles have been distinguished, separated by short periods of quiescence, never longer than 7 years and each one closed by strong eruptions (final eruptions). Each cycle was characterized by frequent eruptions, mainly effusive. The eruption of 1906 (final eruption) represents the strongest manifestation of the Vesuvius activity. The eruption of the 1944, a terminal eruption having both an explosive and effusive character (mixed eruption) was the last one and has signed the passage of the volcano to a state of activity with a closed conduit.

A table reporting the main eruptions related to their age dating and type of volcanic activity has been constructed (Table 1).

3.2. Phlegrean Fields

The Phlegrean Fields are a volcanic field in which many different eruptive centres have been active during the last 39 ky B.P. The geologic history of the Phlegrean Fields has been dominated by two great eruptions: the eruption of the Campanian Ignimbrite (39 ky B.P.) [18] and the eruption of the Neapolitan Yellow Tuff (15 ky B.P.) [140]. These eruptions are connected with two episodes of downthrowing which have generated a complex caldera representing the most evident structure of the Phlegrean volcanic district. This district includes the Phlegrean Fields which are part of the Naples town the Ischia and Procida volcanic islands and the north-western sector of the Naples Bay. The volcanic activity of the Phlegrean district is genetically related with the extensional tectonic events allowing for the formation of the graben of the Campania plain, a tectonic depression bounded northwards by the Massico Mt. and southwards by the Sorrento Peninsula.

The starting age of volcanism in the Phlegrean area is not precisely known: sequences of lavas and pyroclastites having an age of 2 My B.P. have been encountered in deep wells drilled between Villa Literno and Parete [118, 8]. In outcrop the oldest volcanic products have an age of 60 ky B.P. and are composed of pyroclastic deposits and remnants of lava domes.

The interpretation of new stratigraphic data both in outcrop and coming from deep wells [2], also in the light of geological, geomorphological, petrological and geophysical data available in literature [134, 72, 111, 116, 117, 51, 57, 99, 25, 10] has allowed to reconstruct in detail the volcano-tectonic history of the Phlegrean caldera. In particular, the outcrop geology has been reconstructed referring to the deposits of the Campanian Ignimbrite (39 ky B.P.) and the Neapolitan Yellow Tuff (15 ky B.P.), which represent useful marker horizons due to their areal distribution and lateral continuity.

Volcanism Oldest than 39 ky B.P.

The rocks oldest than the Campanian Ignimbrite are exposed only along the slopes bounding the Phlegrean Fields and have an alkalitrachytic composition. They include the lava domes of Punta Marmolite (47 ky B.P.) and of Cuma (39 ky B.P.), the pyroclastic deposits called “Tufi di Torre Franco” (older than 42 ky B.P.) and the relict tuff cone of Mt. Grillo. Pyroclastic deposits having the same stratigraphic height have been encountered in drill sites in the Naples town (Poggioreale, Capodimonte, Ponti Rossi, Chiaiano, Secondigliano).

Table 2. Chronology of the eruptive events in the Phlegrean volcanic district

Ischia	Procida	Phlegrean Fields
First eruptive cycle older than the eruption of the Epomeo Green Tuffs (135-100 ky B.P.)	Volcanoes or eruptive centers of Vivara, Fiumicello, Pozzo Vecchio, Terra Murata, all older than the Epomeo Green Tuffs (55 ky B.P.)	Datations are lacking
Second eruptive cycle of Epomeo Green Tuffs-Citara (55-33 ky B.P.)	Breccia Museo cropping out at Punta della Lingua, Procida Island (40-28 ky B.P.)	Eruptive cycle older than the eruption of the Campanian Ignimbrite (37 ky B.P.)
Third eruptive cycle (37-28 ky B.P.)		Eruptive cycle of the Campanian Ignimbrite (37-28 ky B.P.)
Fourth eruptive prehistorical cycle (18-10 ky B.P.)	Solchiaro volcano (18 ky B.P.)	Mainly subaqueous eruptive cycle of the Yellow Tuffs (37-10 ky B.P.), including the Neapolitan Yellow Tuff
Fifth eruptive cycle, from prehistorical to actual (10-0 ky B.P.)		Mainly subarial eruptive cycle (10-8 ky B.P.) including Agnano 1, Minopoli, Baia and Fondi di Baia
		Recent subaerial cycle (4.5 ky B.P.-1538 A.D.) including Cigliano, Agnano 2, Senga, Astroni, Solfatara and Monte Nuovo.

Campanian Ignimbrite (39 ky B.P.)

The Campanian Ignimbrite is the product of the major explosive eruption happened in the Mediterranean area during the last 200 ky. This eruption has buried the most part of Campania under a thin coverage of tuffs. During the eruption a caldera was formed allowing for the downthrowing of a wide area including the Phlegrean Fields, part of the Naples town and part of the Gulfs of Naples and Pozzuoli.

Volcanism between 39 and 15 ky B.P.

The rocks erupted during the time period ranging between the eruptions of the Campanian Ignimbrite and the Neapolitan Yellow Tuff are exposed along the border of the caldera of the Campanian Ignimbrite, in the Naples town and in the Posillipo hill. The eruptive explosive centres were located in the caldera of the Campanian Ignimbrite (Torregaveta, Monticelli, Monte Echia, S. Martino, Capodimonte, Posillipo). The Pentapalummo and Miseno banks (Gulf of Pozzuoli) have been interpreted as pertaining to this period of activity.

Neapolitan Yellow Tuff (15 ky B.P.)

The eruption of the Neapolitan Yellow Tuff is the second for importance in the Campania area. The corresponding deposits are located in the Neapolitan-Phlegrean area and in the Campania Plain since the Apenninic reliefs. The eruption of the Neapolitan Yellow Tuff was accompanied by the formation of a caldera allowing for the downthrowing of an area including the most part of Phlegrean Fields and Pozzuoli Bay.

A table showing the chronology of the main eruptive events of the Ischia-Procida-Phlegrean Fields system has been constructed (Table 2).

3.3. Ischia

Ischia was characterized by several periods of activity and was the site of great explosive eruptions, such as that one of the Epomeo Green Tuffs (55 ky B.P.). The last eruption took place at 1302 D.C. allowing for the formation of the Arso lava flow. The starting age of the volcanic activity in the island is not precisely known, since the oldest rocks dated back have an age of 150 ky B.P. and pertain to a volcanic complex now eroded and covered by the products of the younger activity. The remnants of this complex have been detected in the south-eastern sector of the island. The products of the activity younger than the formation of this complex are composed of small lava domes having a trachytic and phonolitic composition. These domes (Campagnano, M.te di Vezzi, M.te Barano, Punta della Signora, S. Angelo, Punta del Chiarito, Capo Negro, Punta Imperatore, Monte Vico, inlet of the Ischia Castle) are located along the coasts of the island and range in age from 150 to 74 ky B.P. The following period of volcanic activity was characterized by several explosive eruptions having a variable energy separated by periods of quiescence having a different duration and ending with the eruption of the Epomeo Green Tuffs, happened at 55 ky B.P. [86, 37, 38, 147].

Epomeo Green Tuffs (55 ky B.P.)

The eruption, strongly explosive, of the Epomeo Green Tuffs is responsible for the formation of a caldera occupying the central zone of the island. It allowed for the formation of pyroclastic flows, which partly filled the caldera depression, submerged below the sea level and partly covered the emerged areas. The Green Tuff, deposited in a subaqueous environment is now exposed at the Epomeo Mt and is characterized by the typical green colour due to the long contact with the sea waters.

Volcanism Ranging in Age from 55 to 33 ky B.P.)

Three periods of activity have been distinguished during this time interval. Each period was characterized by specific processes of magmatic differentiation. The first period ranges between 55 and 33 ky B.P. After the eruption of the Epomeo Green Tuffs the volcanic activity continued with explosive eruptions since 33 ky B.P. The rocks originated during these eruptions are exposed along the coastal cliffs between S. Angelo and Punta Imperatore Citara and M.te Vico. They are attributed to eruptive centres which were located along the SW and NW margins of the island.

Volcanism Ranging in Age from 28 to 18 ky B.P.)

The volcanic activity of this period started with the eruption of Grotta di Terra, happened at about 28 ky B.P. along the SE coast of the island. The volcanic activity continued in a scattered way since 18 ky B.P. with the emission of trachytic magmas feeding effusive and explosive eruptions and the emplacement of lava flows fall deposits tuff rings and tuff cones. Corresponding rocks are well exposed at Grotta di Terra Grotta del Mavone M.te di Vezzi S. Anna Carta Romana M.te Cotto Campotese Punta Imperatore and S. Angelo.

Volcanism Younger than 10 ky B.P.

The third period of activity happened at about 10 ky B.P. after a relatively long period of quiescence and continued in historical times with a set of eruptions. The youngest one was dated back at about 1302 D.C. and controlled the formation of the Arso lava flow. This period has been characterized by a strong volcanic activity, both effusive and explosive. Most eruptive centres active during this period are located in the depression to the east of the Mt. Epomeo including Selva del Napolitano M. Trippodi Costa Sparaina Cantariello Posta Lubrano M. Rotaro Punta La Scrofa Cafieri S. Alessandro Ischia Porto and Vateliero. Only some centres, such as that ones originating the Zaro lava flow and the Chiarito pyroclastic deposit are located outwards of the described area.

4. Marine Seismic Reflection and Magnetometric Data in the Naples Bay. From Past to Emerging Technologies

Marine seismic reflection and magnetometric data in the Naples Bay are herein discussed with regard to the evolution from past to emerging technologies.

Seismic surveys are commonly performed through the use of seismic sources generating elastic waves derived from the rapid expansion of underwater bubbles [44, 148]. The explosion is described in terms of a volume source that is characterized by a displaced volume as a function of the time, where the shape and the pressure of the seismic source are unknown. By solving the linear wave equation with boundary conditions representing a volume source at the sea bottom, a simple linear relationship is derived between the source volume and the seismic far-field displacement [148]. It is shown that this relationship remains valid as a good approximation for explosive sources. The volume of explosion gas bubbles may be considered as the source volume within the linear theory and is inferred in practice from the bubble period.

A quantitative theory of the seismic signal generated by underwater explosions has been furnished [148]. The first step is the numerical modeling of the behavior of the gas bubble. This is done through the construction of an extended source including the non-linear zone around the explosion and demonstrating that the crucial source parameter, the time derivative of the volume, remains the same. The final step is the analytic calculation of the seismic far field within the theory of the elastic waves. In the first step the experimental volume-pressure relationship of water is used which is for pressure exceeding a few kilobars different from the linear relationship. A diagram reporting the experimental pressure/volume relationship of water has been constructed [44, 89, 148] (Figure 11).

Several technologies can be used in order to produce an acoustic pressure wave into the water such as the free-falling weights chemical explosives piezoelectric or magneto-resistive sources sparkers boomers airguns and waterguns. Each of these sources has a precise signature and wave frequency which can be considered optimal in function of parameters such as the depth and the resolution. A single high-energy spike is commonly produced by the seismic source detected after crossing the portion of the sea bottom involved by our study.

Sparkers [90] and Boomers [75] are among the most important seismic sources used in the Naples Bay, giving an excellent acoustic response in the detection of subsurface strata. The Sparker is a sound source having a broad spectrum. It is based on an electric spark discharged under water as a sound source in order to study frequency-dependent effects. The output of the sparker is used over the range from 0.1 to 50 kc and offers the possibility of studying in a more detailed way a previously studied range of problems using explosives as sound sources. Typical examples are the geological structures at shallow depths below the sea bottom. The combination of the sparker with the recorder is a powerful method for observing both short and long-period variations in these phenomena. The sparker sound pulse is emitted when the helix starts its sweep near the edge of the paper and the sound received over various paths is recorded as it arrives at the hydrophone. Sparker geophysical experiments [90] have shown that the first arrival noted at the shorter ranges is directly transmitted through the water and it is double owing to the bubble pulse from the spark. The second group of arrivals is a single bottom reflection which is somewhat complicated, the principal features being four arrivals near its leading edge. Two of the arrivals are the initial sound and bubble pulse, while the others are related to them by travel-time differences to be expected from bottom-reflected travel paths directly from the sparker and, after one reflection, from the surface near the spark. The third group of arrivals corresponds to paths involving two reflections from the bottom; they are seen to be weakly developed over most of the experiment. The tests have confirmed that the sparker source, giving good results, is too weak for any other than short range phenomena.

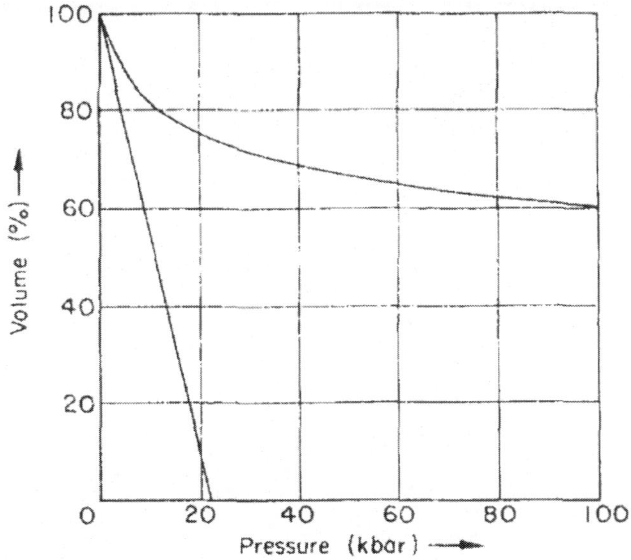

Figure 11. Diagram reporting the experimental pressure/volume relationship of water (modified after Cole, 1965; Kennedy and Holser, 1966; Wielandt, 1975).

The boomer sonar source has been deeply described in seismic profiling [75]. It is an electromagnetically driven sound source consisting of a triggered capacitor-bank that discharges through a flat coil or coils. Eddy currents are induced in aluminium plates that are held against the coil by heavy springs or rubber bumpers. The plates are repelled and a cavitation volume is created into the water, acting as a source of low frequency sound. The boomer has been used as the transmitter of a continuous seismic profiling system at sea. The boomer is a combination of a transducer of an electronic power supply of a capacitor storage bank and of a triggering device. The system has been repeatedly modified to increase its acoustic output and to improve its efficience and reliability. The sound producing element of the boomer is a flat coil of wire which is magnetically coupled to either one, or in later models, two aluminium plates. The energy stored in electrical capacitors is discharged into the coil, causing induction currents in the plates that result in an outward force. Seismic profiles have been recorded in many locations with the 1000 joule-unit, in both salt and fresh waters. Echoes from interfaces 300 meters below the bottom in shallow water have been detected and recorded.

The Multispot Extended Array Sparker (M.E.A.S.) is a seismic source consisting of sparker electrodes disposed on a square metal cage. This kind of system, patented by the Institute of Oceanology of Istituto Universitario Navale of Naples (Italy) allows for obtaining a good penetration of the seismic signal coupled with high resolution seismic data. The M.E.A.S. signal is a short impulse with a large frequency spectrum content [102, 11] (Figure 1).

The M.E.A.S. consists of 4x4 meter metallic cage which holds a planar array of 36 equally spaced electrodes, divided into two banks of 18, each powered with 7.5 kJ (Figure 12) [103]. The penetration reaches a depth of 2 seconds (twt) in soft sediments. The resolution of the seismic source is 6 meters, but the ghost generated by the ship's hull may degrade the vertical resolution down to 25 meters. The firing rate is 6 seconds.

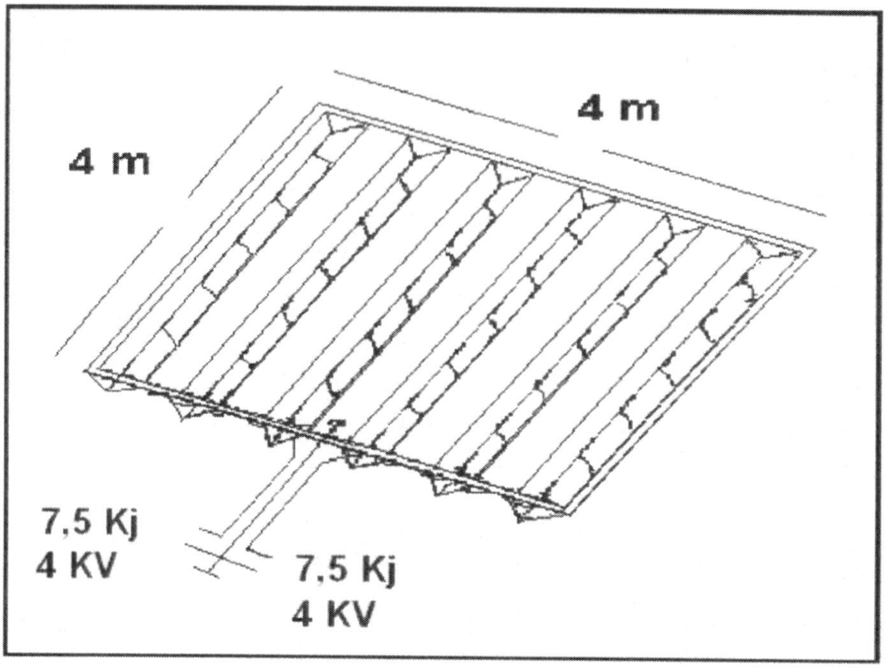

Figure 12. MEAS – Multiple electrode Extended Array Sparker (modified after Mirabile et al., 2000).

Mirabile et al. (2000) [103] have obtained interesting geologic results through the use of M.E.A.S. on the Tyrrhenian margin south of the Gulf of Naples the north-western Tyrrhenian margin the western Naples Gulf and Phlegrean Fields and the geothermal resources in the Pozzuoli Gulf. In the first area the Ischia volcano with a radius of 20 kilometers at the sea bottom is the widest one. Multichannel seismic and marine magnetic data [103] have evidenced the occurrence of Mesozoic tilted blocks related to Sorrento Peninsula covered by thick siliciclastic units whose top is represented by a Miocene unconformity.

Sub-horizontal onlapping sediments, Pliocenic-Holocenic in age, cover the unconformity. A significant volcanic intrusion is interstratified with the aforementioned units and is not constrained by magnetic data. A tilted limestone block is bounded by the Naples Bay canyons, supposed to be tectonically-controlled.

A detailed magnetic map (Figure 13) [103]. has been collected by the Istituto Universitario Navale of Naples, Italy, suggesting the occurrence of previously unknown underwater volcanoes [103]. Mirabile et al. (2000) [103]. have suggested the occurrence of a lava layer at depths ranging between 1.5 and 3.0 sec to the WNW of the Procida-Pozzuoli magnetic maximum. Other lava deposits have been associated to the Villa Literno-Parete volcanism related to the tectonic activity of 41st parallel fault system. Moreover, these authors have evidenced the structure of the carbonate basement characterized by structural highs and tilted blocks at increasing depths. The strong downthrowing of the Phlegrean Fields area is confirmed by the seismic markers referred to the top of the Miocene unconformity and to the top of carbonate platforms. The Phlegrean volcanism is supposed to be originated from the prolongation of the Magnaghi canyon, bounded to the north by the Aloha bank.

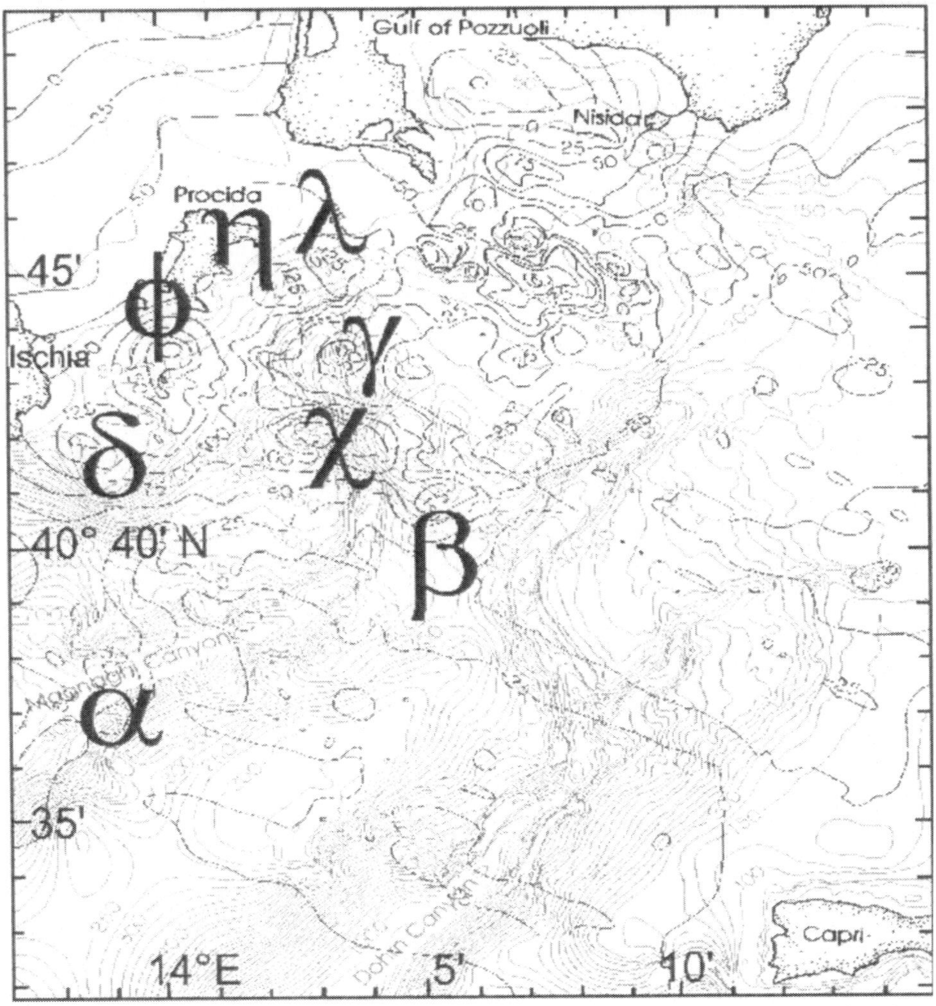

Figure 13. Sea beam bathymetry (gray lines) and magnetic anomalies (black) of a sector of the Naples Bay (modified after Mirabile et al., 2000). Greek letters indicate submarine volcanoes detected by Mirabile et al. (2000).

Regarding the geothermal resources several flat spots have been found indicating the presence of accumulated fluids in the upper sequences of the Pozzuoli Gulf. Some important fluid traps have been detected. The shallow reservoirs do not justify the intense and deep presence of vapour and gas in the volcanic sediments.

By continuing with the seismic technologies in the Gulf of Naples, a more recent technology is represented by the Sparker source SAM, characterized by a varying number of electrodes that can be disposed as "dual in line" (SAM96) and "planar array" multi-electrode electro-acoustic source (SAM 400/800) [45]. The SAM 96 seismic source has been developed in the frame of the PNRA Project [45], but has been employed also in the CARG Project of the Campania region [23, 45, 10, 12, 4].

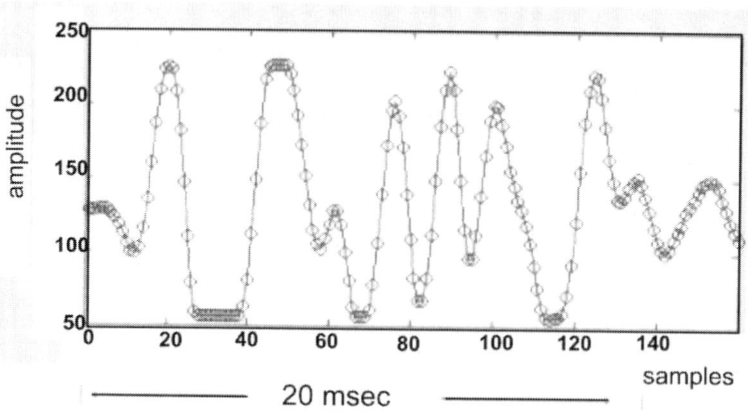

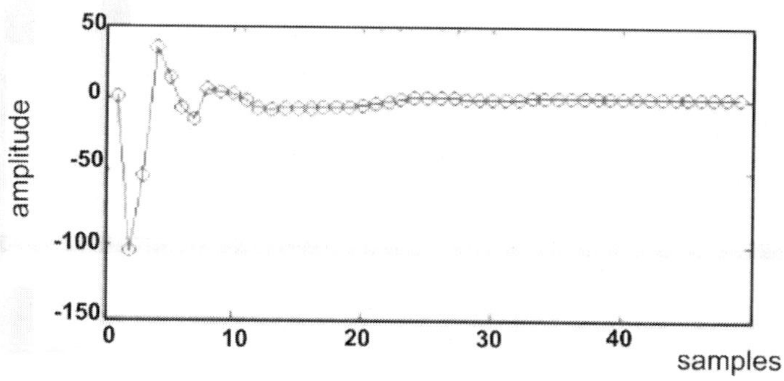

Figure 14. Sketch diagram showing a comparison between the direct wave coming from traditional 1 kJ sparker source to geophones array (sampled at 8000 samples/sec) and the direct wave coming from the Multi-tip (SAM96) sparker source (400 J) to geophones array (sampled at 8000 samples/sec).

A sketch diagram has been constructed (Figure 14) showing the direct wave coming from traditional 1 kJ sparker source to geophones array (sampled at 8000 samples/sec) compared with the direct wave coming from the Multi-tip (SAM96) sparker source (400 J) to geophones array (sampled at 8000 samples/sec).

Among the most used seismic sources in the Naples Bay there are the waterguns and the airguns, mainly adopted in research programs on marine geological mapping [6, 8, 9, 51]. The watergun has been designed by the Society for the Development of the Applied Research for high resolution seismic profiling. It produces an acoustic output from the sudden collapse of a cavitation volume in the water [88]. The Figure 15 [88] represents a watergun at three stages of its firing cycle. At the beginning of the watergun cycle, the compressed air acts on the central piston moving it downwards and contemporaneously expelling the water contained in the cylinder ahead the piston. After the piston is arrested, the moving plugs of the water are separated from the piston causing the formation of a cavitation volume. It produces a seismic output linear to the kinetic energy of the water plug. Consequently, the air is vented from the upper chamber and the hydrostatic pressure pushes the piston to its upstroke position triggering for the next firing cycle. Excellent examples of Watergun profiles recorded by the CNR-IAMC of Naples during seismic and magnetic survey of the Naples Bay have been shown by Aiello et al. (2005) [6]. The airgun is based on the discharge of compressed air into water, producing short seismic signals characterized by high energy [88]. The airgun is composed of two pressure chambers sealed by a double ended piston, which has an axial opening to allow for the passage of air when the chambers are closed and immerged into the water. The airgun is fired by opening the solenoid valve; the piston rapidly moves, discharging the high pressure air from the lower chamber into the water. The pressure difference causes the piston to rapidly return to its initial position. High-pressure air begins to enter the lower chamber and the firing cycle is repeated. Good examples of Airgun profiles recorded by the CNR-IAMC of Naples during seismic survey of the Naples Bay have been shown by D'Argenio et al. (2004) [51].

Seismic and magnetometric data have been significantly compared in the Naples Bay [6]. A densely spaced grid of magnetic profiles coupled with multichannel seismics has been acquired by the CNR-IAMC of Naples in the Naples area onboard of the R/V Urania. Shallow volcanic structures in the subsurface of the gulf have been recognized through seismo-stratigraphic analysis of seismic profiles. The volcanic nature of some of these structures has been inferred through the identification of the magnetic anomalies on the high resolution magnetic anomaly map of the Naples Bay.

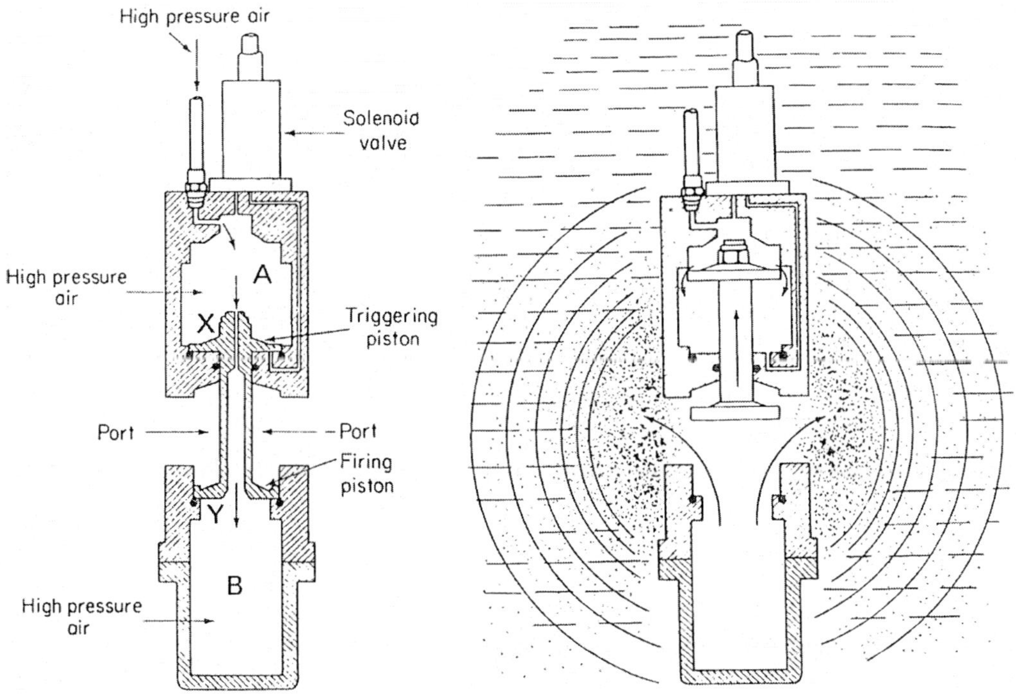

Figure 15. Firing cycle of a watergun seismic source, consisting of three stages (modified after Jones, 1999).

Magnetic data have been recorded by using the G811 proton magnetometer. The sensor was placed in a towed fish generally at 200 m from the ship and 15 m b.s.l. The depth of the magnetometer was regularly controlled and recorded. The cruising speed did not exceed 6 knots. The data were sampled at 3 sec, corresponding to an average sampling rate of about 6.25 m. An accurate magnetic data processing has been performed to preserve the data information contents. Initially, the raw data were edited manually or using a non-linear filter to remove the spikes controlled by non-geological sources. The marine paths were repositioned taking into account the offset distance between the fish and the GPS positioning system. The elimination of the diurnal component was carefully controlled adopting as base magnetic station the geomagnetic observatory of L'Aquila (42°N, 12°E) kindly supplied by the Istituto Nazionale di Geofisica e Vulcanologia (INGV, Rome, Italy). The measuring periods were characterized by a quiet magnetic activity. An evident correlation of the spectral contents of the two datasets for the periods relative to these variations has been observed.

The values of the main magnetic field must be subtracted by the measured values corrected as previously explained in order to recognize the total field anomaly linked to the geological structures. The adopted model in magnetic data processing presented in this chapter is the Italian Geomagnetic Reference Field (ItGRF) with values calculated for the reduction epoch, year 2000 [46, 5]. Some factors contribute to explain the shifts in the magnetic field among adjacent navigation lines. These include the differences in level among the navigation lines and the inadequate removal of the magnetic field variations for the diurnal component. The magnetic data have been elaborated according to a procedure of leveling consisting of the removal of the short period magnetic variations. A histogram showing the distribution of mistakes after the several phases of magnetic processing is shown in Figure 16 [5], shown on the right in the diagram.

5. RESULTS

5.1. Somma-Vesuvius

An integrated interpretation of seismic and magnetic profiles recorded offshore the Somma-Vesuvius volcano has been carried out [7]. A large volcanic structure located offshore the Torre del Greco town and representing the seawards elongment of the Vesuvius structure has been modelled through comparison of seismic and magnetic data.

Late Quaternary stratigraphic architecture of the Naples Bay is controlled by the interactions between volcanic and sedimentary processes. Rapid lateral variations between acoustically-transparent seismic units occur. These units have been interpreted as volcanic deposits erupted during the Vesuvian and Phlegrean eruptive activities interlayered with Pleistocene prograding wedges [51].

In the study area the magnetic properties of rocks and sediments have allowed to discriminate the volcanic nature of subsurface intrusions evident on seismic profiles. A semi-quantitative integrated interpretation of bathymetric and seismic data has been performed resulting in a detailed topographic and seismic reconstruction of the Torre del Greco volcanic structure.

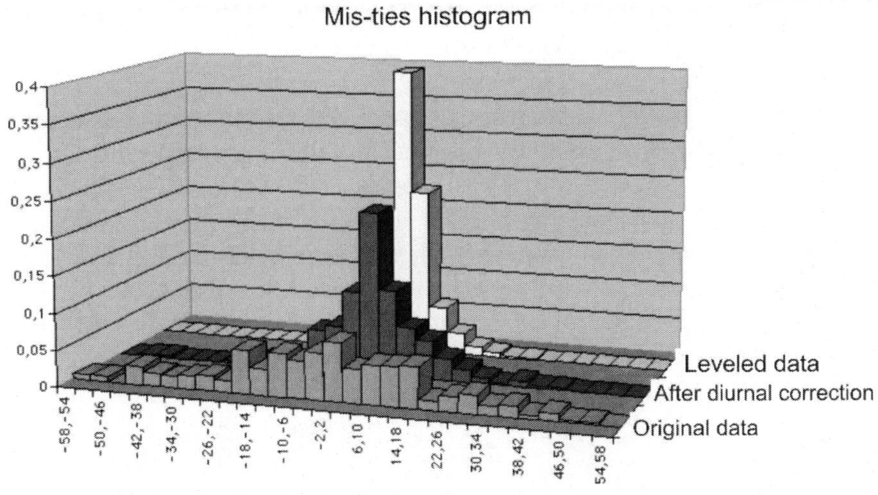

Figure 16. Histogram showing the mistake distribution after several measurements of magnetic processing (modified after Aiello et al., 2004).

The Vesuvius volcano has been studied in detail. Vesuvius monitoring and knowledge with respect to the state of the art and perspectives have been recently resumed [13, 15, 22, 30, 41, 48, 49, 49, 52, 55, 56, 71, 76, 79, 84, 113, 125, 138].

Alessio et al. (2013) [13] have studied the flood hazard of the Somma-Vesuvius region based on historical and geomorphological data and have constructed a susceptibility map of the volcano. High resolution DEMs and detailed geomorphological maps have been analyzed and processed in a GIS environment carrying out a comparative study of present and past morphologies of the volcanic edifice including changes of infrastructures and buildings. These results together with historical data and rainfall data for all flood events have been processed for the individuation of the drainage basin areas of Somma-Vesuvius, where the flood phenomena must be more probable in the future.

Amoruso and Crescentini (2013) [15] have investigated the analytical models of volcanic ellipsoidal expansion sources. The modeling of earthquakes and surficial deformation in volcanic and geothermal areas involves the expansion sources. Given an ensemble of ellipsoidal or tensile expansion sources and double-couple ones it is important to obtain the equivalent single moment tensor representation. The authors focused on the volume change estimate in the case of single sources (finite pressurized ellipsoidal sources) presenting the expressions for the computation of the volume change and surficial displacement in a closed analytical form.

Berrino et al. (2013) [22] have precised that thirty years of precise gravity measurements at Mt. Vesuvius represent an approach to detect underground mass movements. On Mt. Vesuvius the gravity network consists of selected sites coinciding with leveling benchmarks to remove the effects of the elevation changes from gravity variations. The reference station is located in Naples, outside the volcanic area. Moderate gravity changes on short times have been generally observed. On long-term significant gravity changes occurred and the overall fields displayed well-defined patterns. The gravity changes appear related with the seismic crises and with changes of tidal parameters obtained by continuous measurements. The absence of significant ground deformation implies masses redistribution, essentially density changes, such as fluid migration at the depth of the seismic foci. The fluid migration may occur through pre-existing geological structures and through new fractures generated by seismic activity.

Capuano et al. (2013) [30] have studied the P-wave velocity and density structure beneath Vesuvius suggesting the occurrence of a magma body in the upper edifice. A high resolution image of the compressional wave velocity and density structure in the shallow edifice has been derived from the simultaneous inversion of travel times and hypocentral parameters of local earthquakes and from gravity inversion. The tomography has been improved by adding to the earthquake data a set of land based shots, used for constraining the travel time residuals. Obtained results have given a high resolution image of the P-wave velocity structure with details down to 300-500 m. The local seismicity appears to extend down 5 km below the central crater, distributed into two clusters and separated by an anomalously high V_P region positioned at around 1 km of depth. The seismicity is located in vertical cracks and delimited by a high velocity body located along past intrusive body corresponding to Mt. Somma.

Cioni et al. (2013) [41] have presented new volcanological data on the timing of the upward migration of the post A.D. 79 magma chamber. The authors have presented new data on the deposits of the Santa Maria Member, the eruption cycle occurred at Vesuvius in the period between the 79 D.C. eruption and the A.D. 472 eruption. Stratigraphic studies and component analyses have suggested that the activity was characterized by mixed hydromagmatic and magmatic processes. The eruption style has been interpreted as repeated alternations of continuous and prolonged ash emission activity interlayered with strombolian phases. During the eruption cycle the magma maintained a phonotephritic composition.

Cubellis and Marturano (2013) [48] have evaluated the felt index, the source parameters and the ground motion for earthquakes at Vesuvius volcano. The results of non-instrumental surveys have been re-taken to purpose new analyses regarding the source mechanisms and the causative faults. The earthquake of 9 October 1999 has been evaluated, the most intense one after the 1944 event. The intensity was evaluated by using integer values of the MCS scale and the felt index as a continuous parameter. The values of magnitude and attenuation determinated by applying macroseismic models to the data and compared to the instrumental ones have been used to assess the size of the historical Vesuvius earthquake.

Cusano et al. (2013) [49] have investigated the first long period earthquake detected in the background seismicity at the Mt. Vesuvius. Typical earthquakes occurring at the Vesuvius are volcano-tectonic in origin. On July 20 2003 an earthquake with low and narrow frequency content has been detected. The seismograms presented an emerged onset and a monochromatic spectrum at all stations. The event was located close to the crater axis and an equivalent duration magnitude of 0.6 has been estimated. The nature of this event has been investigated by comparing its features with those of two volcano-tectonic earthquakes occurred inside the same rock volume.

D'Auria et al. (2013) [52] have analyzed the recent seismicity of the Vesuvius (1999-2012). Since the 1972 a continuous observation with electromagnetic seismometers allowed for the compilation of a detailed earthquake catalogue. The Gutenberg-Richter distribution of the magnitudes has shown a temporal decrease of the b-value since the 1985, with current values close to 1.0. The temporal pattern of strain release shows a non-stationary behavior with periods of increased release rates. The spatial distribution of the seismicity consists of two main seismogenic volumes, one with hypocenters clustered below the Mt Vesuvius crater at depths between 1 and 6 kilometres. The statistical properties of the seismicity occurring between the two volumes and their spatial and temporal patterns have been evaluated. The obtained results have pointed to gravity-induced stresses as the source of the shallow seismicity.

Del Pezzo and Bianco (2013) [55] have shown a new method to look at the seismic velocity and attenuation tomographic imaging. New velocity and attenuation images of the geological structures below the Mt. Vesuvius have been obtained using the programming facilities as well as the enhanced graphical power of the software Mathematica. The velocity and attenuation space distribution, already calculated inverting respectively P-wave travel times and amplitude spectra of local VT quakes are first optimally interpolated and then graphically represented. The method is particularly adopted to represent attenuation images where the space variations of the parameters are strong with respect to their average.

Di Maio et al. (2013) [71] have examined the electric effects induced by artificial seismic sources at Somma-Vesuvius volcano. Self-potential measurements at Somma-Vesuvius have been acquired in conjuction with an active seismic tomographic survey. This study has provided further confirmation to the occurrence of seismo-electric coupling and to the identification of sites suitable for self-potential signal monitoring. The data have shown significant changes in the natural electric field pattern. These variations were observed after explosions occurred at a distance less than 5 km from the SP dipole arrays. The NW-SE component of the natural electric field was more sensible to the shots than the NE-SW one and the major effects did not correspond to the nearest shots. Such evidences were interpreted considering the underground electrical properties, as deduced by previous detailed resistivity and self-potential surveys performed in the study area.

Esposito et al. (2013) [76] have shown that the neural analysis of seismic data may be applied to the monitoring of the Vesuvius volcano. Among the existing methods the neural analysis may be considered as an efficient tool for detection and discrimination and may be integrated into intelligent systems for the automatic classification of seismic events. A technique has been applied for the analysis and classification of seismic signals in order to improve the automatic event detection. The examined dataset includes 1500 records divided into four typologies of events: earthquakes, landslides, artificial explosions and other. The clustering has been obtained using a Self-Organizing Map neural network, not requiring a classification of the seismic signal. The SOM map has been separated into different areas, each containing the events of a defined type.

Federico et al. (2013) [79] have discussed the groundwater geochemistry of the Vesuvius, highlighting some implications for the volcano surveillance and the relationships with hydrological and seismic signals. Water samples have been collected in a permanent network of wells and springs located in the areas mostly affected by the ascent of magmatic volatiles. Since the geochemical parameters describe the behavior of groundwater at the Vesuvius, the time distribution of earthquakes has been examined. The water chemistry and seismological datasets are characterized by a good agreement. The termination of the phase of intense deep degassing is associated with a change in water chemistry and a peculiar seismic event, recorded in July 2003. The seismic and geochemical patterns are interpreted accordingly to temporal variations in the regional and local stress field.

Local and regional earthquakes recorded at Vesuvius in 1997 and 1998 by a temporary seismic array have been analyzed to estimate the slowness and backazimuth of correlated phases [84]. The backazimuth observed for the directed P-wave has been compared with the expected direction. A wavefront distortion produced by the propagation through a non-homogeneous velocity structure has been shown. Two different velocity models have been obtained applying a finite-difference method.

Orazi et al. (2013) [113] have discussed the seismic monitoring of Vesuvius. Its activity is currently characterized by moderate seismicity, with hypocenters located beneath the crater zone with depth rarely exceeding 5 km and magnitudes generally less than 3. The current configuration of the seismic monitoring network of Mt. Vesuvius consists of 18 seismic stations and 7 infrasound microphones. During the period 2006-2010 a seismic array with 48 channels was also operative. The station distribution has provided an appropriate coverage of the area around the volcanic edifice. The current development of the network and its geometry under conditions of low seismic noise has allowed for the location of seismic events around the volcanic edifice.

A detailed processing of high resolution magnetic data recorded by the CNR-IAMC of Naples offshore the Vesuvius volcano has enabled the identification of weak magnetic anomalies sub-circular in shape and aligned in the NNW-SSE direction located in the submerged part of the Vesuvius volcano [141, 5, 6].

An aeromagnetic survey of the Somma-Vesuvius volcanic area completed by Paoletti et al. (2005) [120] later showed that the field map is dominated by the large anomaly related to the Vesuvius itself and characterized by a roughly elliptical shape (Figure 5). Other important magnetic features evident from the map of Figure 5 [120] are located both at the base of the edifice and in the north-eastern area of the edifice. In the first sector they consist of a narrow anomaly on the western slope of the edifice and of small anomalies located on the south-eastern slope of the volcano. In the second sector an elliptical anomaly has been noticed and a general radial trend of the magnetic field has been observed (Figure 5).

A comparison between the new aeromagnetic data [120] and the magnetic data acquired on Vesuvius by AGIP oil company in 1981 [34] has shown that the field map is rather more complex. The Agip map did not show the magnetic anomalies located onshore along the slopes of the volcano. On the other side, the Agip map shows the anomalies located in the southern part of the edifice. Both the magnetic datasets confirm the NW-SE magnetic trend from the Vesuvius edifice through the towns of Boscotrecase and Boscoreale and the W-E trend through the towns of Torre Annunziata and Pompei.

A first interpretation of the magnetic dataset in terms of geological structure has been permitted by the maps of the analytic signal and horizontal derivative [120], allowing for the location of the magnetic lateral boundaries. They have revealed the occurrence of small parasitic vents. Some magnetic maxima in the volcanic edifice south of the crater seem to correspond with the caldera rim suggested by previous scientific literature. Three main linear trends have been observed: a NE-SW one, a NW-SE one and a W-E one.

Based on the same marine magnetic dataset presented in this chapter a high resolution magnetic anomaly map of the Naples Bay has been constructed [5]. A high detail was ensured by a more advanced spatial coverage of the magnetic data if compared with previous magnetic measurements such as airborne magnetic survey [3, 31]. Two main belts of sharp and delineated magnetic anomalies have been identified in the Naples Bay [5]. The first belt (discussed in the present chapter) is located offshore the Vesuvius volcano, while the second belt is located offshore the Phlegrean caldera.

The seismic stratigraphy offshore the Vesuvius volcanic complex has been studied in detail [7]. A sketch table reporting the main seismic units identified offshore the Vesuvius volcano has been constructed (see the following table).

The multichannel seismic and magnetic profile GRNA09 is discussed in this chapter (Figure 17) [7]. The correlation among seismic and magnetic profile has shown that three main elevated peaks of a large volcanic structure namely the "Torre del Greco volcanic structure" (due to its location offshore the town) are related to three magnetic anomaly maxima having intensities up to 400 nT. The volcanic structure overlies the CI unit (Table 3). It appears eroded at its top, probably in a subaerial environment and downthrown by normal faults. The vent is overlain by Late Pleistocene marine and coastal deposits and then by the Holocene mud wedge (Table 3). The occurrence of these units has evidenced that the volcanic structure has been fossilized by recent sedimentation.

A map of the distribution of the volcanic structures superimposed to the magnetic anomaly map of the study area has been constructed (Figure 18) [7]. Based on a high-resolution magnetic survey of the Naples Bay presented in this chapter [5, 142, 141] sharp and delineated magnetic anomalies have been recognized in the eastern sector of the Naples Bay, offshore the Somma-Vesuvius volcano. These anomalies have not distinguished in previous magnetic studies on the Naples Bay [3, 36] due to the larger grid spacing of magnetic and/or aeromagnetic lines.

Table 3. Seismic units offshore the Somma-Vesuvius volcano

Seismic unit	Seismic facies	Geologic interpretation
E unit	Parallel and continuous seismic reflectors.	Holocene mud wedge
D unit	Deepest seismic unit, characterized by low angle slightly dipping reflectors and indicating a progradation axis with a NW-SE orientation. The unit is truncated by an erosional unconformity marking also the base of the CI unit.	Upper part of Middle-Late Pleistocene prograding wedge supplied by the Sarno river mouth
CI unit	Important seismic unit, characterized by acoustically transparent seismic unit and wedge-shaped external geometry. Stratigraphic marker suggesting that both the isolated bodies (BV) and the Torre del Greco volcanic structure (BV unit) must be younger than 37 ky B.P., which is the age dating of the CI unit.	Campanian Ignimbrite pyroclastic flow deposits (37 ky B.P.) widely occurring in the subsurface of the eastern Naples Bay. The CI deposits filled the whole Campania Plain during a major eruption related to the activity of the Phlegrean Fields volcanic complex.
BV unit	Volcanic seismic units, acoustically-transparent, having a tabular external geometry.	Submerged and or buried volcanic vents genetically related to the Vesuvius volcano and including the Torre del Greco structure.
B unit	Volcanic seismic units, acoustically-transparent having a mounded external geometry.	Dome-shaped, isolated and buried volcanic structures.

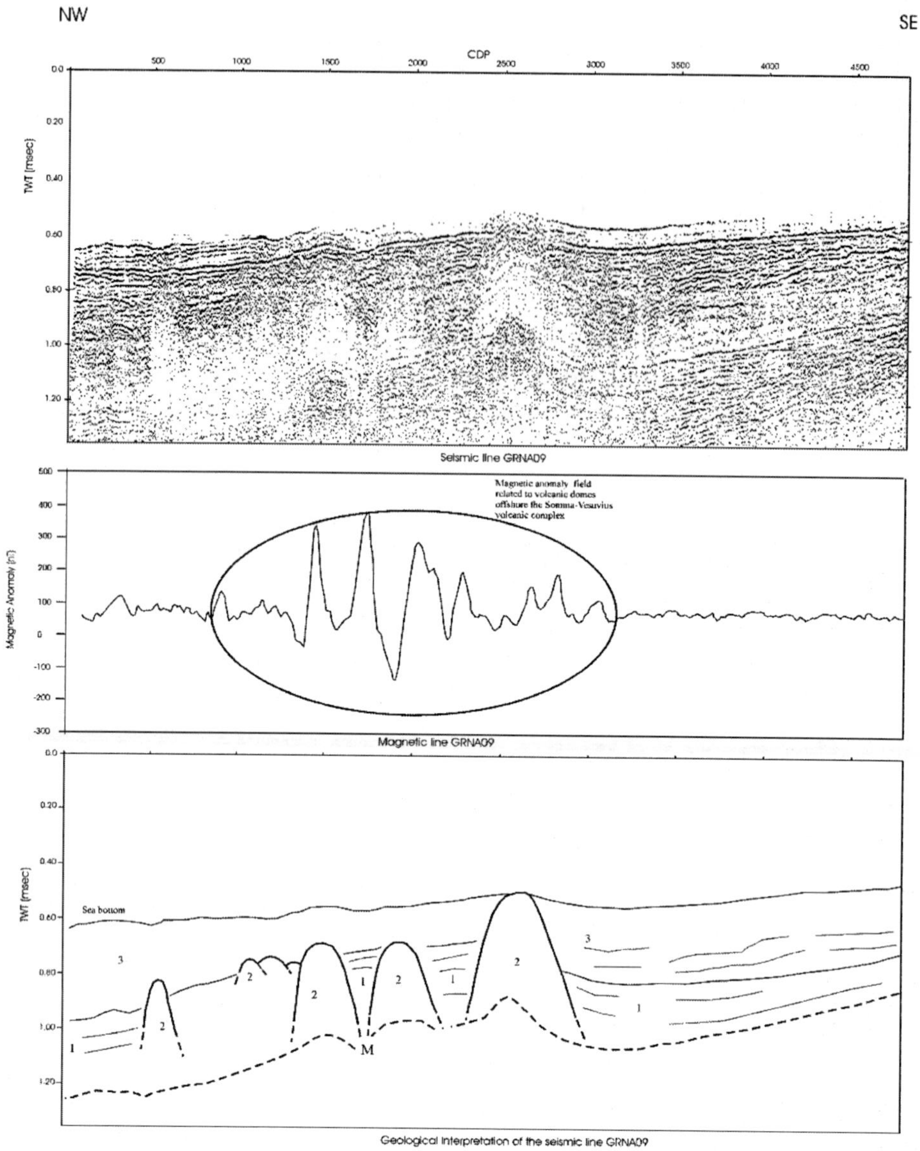

Figure 17. Multichannel seismic profile GRNA09, magnetic line GRNA09 and corresponding geologic interpretation (modified after Aiello et al., 2010). Key. 1: Campanian Ignimbrite (CI seismic unit). 2: submerged and/or buried volcanic vents genetically related to the Somma-Vesuvius volcanic complex. 3: Late Pleistocene-Holocene marine sediments.

The total magnetic field offshore the Somma-Vesuvius volcanic complex (Figure 4) shows that the shape of the anomaly is dipolar. There is no apparent effect caused by the occurrence of remnant magnetization with respect to that one of the present-day main field. Three main anomalies are evident following a NW-SE direction. The most southern one is the less intense, located in a relatively quiet magnetic area. Some circular structures, aligned NNW-SSE occur in the Torre del Greco offshore. Other small and localized anomalies, less intense than the aforementioned ones are located in the Torre Annunziata offshore.

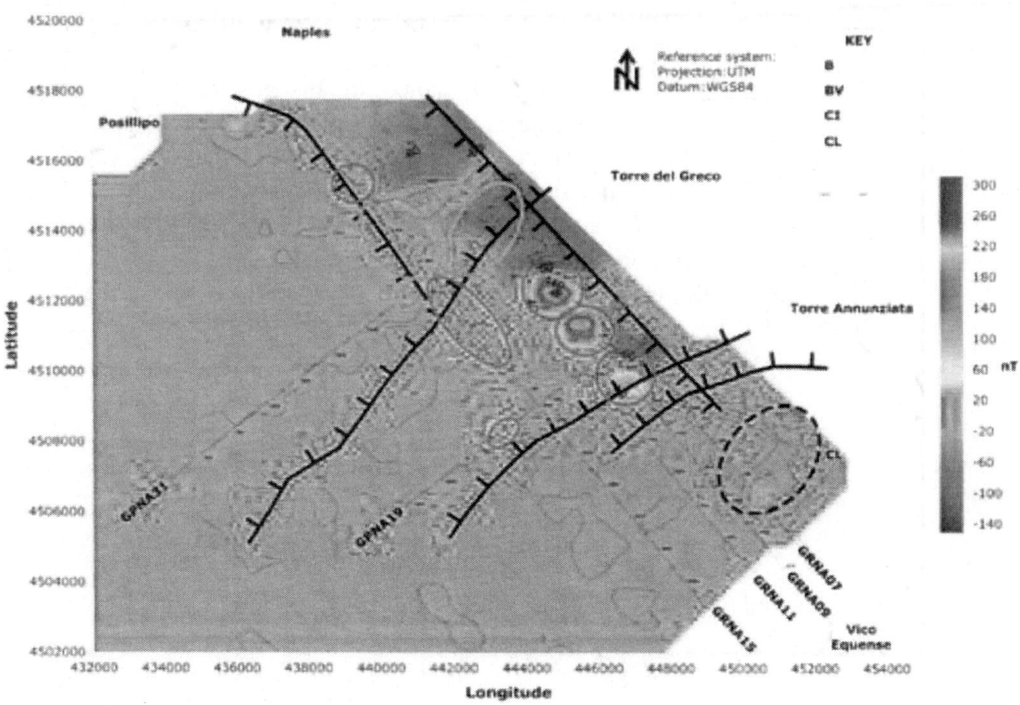

Figure 18. Map of the distribution of the volcanic structures superimposed to a magnetic anomaly map of the study area (modified after Aiello et al., 2010). Key. B: dome-shaped, isolated and buried volcanic structures, not related with magnetic anomalies. BV: submerged and/or buried volcanic vents genetically related to the Somma-Vesuvius volcanic complex (Torre del Greco structure), corresponding to a NNW-SSE trending magnetic anomaly field. CI: Campanian Ignimbrite seismic unit, not correlated with magnetic anomalies. CL: lava flows of the Somma-Vesuvius volcanic complex, corresponding to slight magnetic anomalies offshore the Torre Annunziata town. Normal faults recognized through geological interpretation of seismic profiles have also been represented.

The geological structures identified through seismic interpretation (see Table 3) are located in a complex magnetic anomaly area which is made up of several anomalies reaching a maximum intensity of 400 nT. This area represents the offshore prolongation of the Vesuvius volcano. Main trend of magnetic anomalies and related seismic structures suggests the occurrence of a NNW-SSE trending structural lineament. This is contrary to what suggested by the previous papers, in which systems of NE-SW trending normal faults have been highlighted.

5.2. Phlegrean Fields and Gulf of Pozzuoli

Marine geological setting of the Gulf of Pozzuoli has been relatively poorly studied. Among the most interesting studies there are the papers of Pescatore et al. (1984) [124] and De Pippo et al. (1984) [63]. During two oceanographic cruises undertaken in 1983 in the Gulf of Pozzuoli acoustic profiles were collected (EDO 3.5 kHz, Surfboom and Sparker 6 kJ) [124]. The profiles have evidenced the occurrence of three seismic units separated by two main unconformities. The first unconformity, found only in the inner part of the gulf has been correlated with an important volcano-tectonic event dated back 6000-5000 years from recent times. This event has triggered a strong tectonic uplift, confirmed by the occurrence of the marine terrace of "La Starza," found at about 30 meters on the present-day sea level, on which the town of Pozzuoli is located. The second unconformity is referred to the Wurmian regression and is represented by a continuous surface in the outer continental shelf. The distinguished units constitute a single tectonic structure limited by two important faults. The first one corresponds to the edge of the outer platform and the second one, having a E-W trending is located onshore to the north of Pozzuoli. The Gulf of Pozzuoli represents perhaps an area of recent subsidence from 12 ky B.P. to recent times. The physiographic features of the Gulf of Pozzuoli have been investigated through EDO 3.5 kHz profiles and Sidescan Sonar lines [63]. Four morphological units have been distinguished. A coastal terrace, extending up to 50 m of water; a sub-circular central basin, having average water depths of 100 m and representing the depocenter of the Holocene sedimentation; a belt of submarine volcanic banks, represented by the Miseno, Nisida and Pentapalummo banks and finally, a continental shelf genetically related to the last sea level lowstand, extending from the volcanic banks, reaching water depths of about 165 meters and undergoing recent subsidence. The central basin has been originated by the volcanic collapse of the Wurmian shelf, while the tectonic uplift has involved the Pozzuoli area.

A grid of Sparker Multitip seismic profiles acquired in the Gulf of Pozzuoli in the frame of research projects of submarine geological mapping has been interpreted adding new tectono-stratigraphic evidence on submarine prolongation of the Phlegrean caldera [10].

A sketch table of the seismic units in the Gulf of Pozzuoli has been constructed for this chapter (Table 4).

The interpretation of seismo-stratigraphic data in the Gulf of Pozzuoli is simplified by the seismic profile L69_07 (Figure 19) [10]. The seismic line runs between the Nisida inlet (Naples) and the western continental shelf of the Gulf of Pozzuoli. Eight seismic units have been distinguished through the geological interpretation of seismic profile. The V3 volcaniclastic unit is the deepest unit and appears deformed in anticlinalic structures (Nisida and Pozzuoli anticlines). It is overlain by the G3 sedimentary unit, representing the lower sedimentary unit of the Pozzuoli basin fill. A wide palaeo-landslide, namely the lsl1 unconformably overlies the V3 and G3 units proceeding towards Pozzuoli. The lsl1 unit shows facies hetheropy into the LST and TST units proceeding towards Naples.

Marine magnetic data of the Gulf of Pozzuoli have been studied in detail [10, 11]. Marine magnetic measurements have been discussed in detail by Nabighian et al. (2005) [109]. They led to the development of the sea floor spreading model based on the concept that the sea floor is magnetized either positively or negatively depending on the polarity epoch of the Earth's magnetic field. New sea floor is created at mid-ocean ridges and becomes part of oceanic plates moving away from the spreading center. The magnetic anomalies along a section perpendicular to the spreading center show a regular pattern of highs and lows, often symmetric about the spreading center that can be calibrated in age to the geomagnetic time scale.

Table 4. Table of the seismic units in the Gulf of Pozzuoli

Seismic unit	Seismic facies	Geologic interpretation	Location
HST	Progradational to parallel seismic reflectors	Highstand system tract of the Late Quaternary depositional sequence	Eastern Pozzuoli Gulf
TST	Retrogradational seismic reflectors	Transgressive system tract of the Late Quaternary depositional sequence	Eastern Pozzuoli Gulf
Lsl	Wedge-shaped, chaotic to discontinuous seismic unit	Landslide deposits intercalated in the upper part of the Lowstand System Tract; locally occurrence of palaeochannels.	Gulf of Pozzuoli, south of Miseno Cape.
LST	Progradational seismic reflectors, erosionally truncated at their top.	Lowstand system tract of the Late Quaternary depositional sequence	Inner continental shelf off Pozzuoli town
G1	Parallel and continuous seismic reflectors	Upper sedimentary unit of the basin fill, attaining maximum thickness in the depocenter of the central basin.	Gulf of Pozzuoli
lsl2	Wedge-shaped, chaotic to discontinuous seismic unit	Fossil landslide overlying the G2 marine unit and underlying the Lowstand System Tract of the Late Quaternary depositional sequence	Eastern Pozzuoli Gulf
NYT/PC	NYT: wedge-shaped acoustically-transparent volcanic seismic unit PC: mound-shaped, acoustically transparent volcanic bodies interlayered with parallel reflectors.	NYT: pyroclastic deposits of the Neapolitan Yellow Tuff (12 ka) deposited in Naples and Pozzuoli offshore. PC: Tuff cones of the Nisida volcanic complex in facies hetheropy with the Neapolitan Yellow Tuffs and interstratified with the G3 marine deposits.	Naples and Pozzuoli offshore.
G2	Parallel and continuous seismic reflectors.	Intermediate sedimentary unit of the basin fill, probably composed of clastic deposits; deposited in the	Pozzuoli Gulf
		whole Pozzuoli Gulf; strongly involved by wedging and growth	
lsl1	Wedge-shaped, chaotic to discontinuous seismic unit	Wide palaeo-landslide overlying the V3 volcaniclastic unit and coeval with the basal part of the G2 marine unit.	Eastern Pozzuoli Gulf
pyr2	Continuous progradational to parallel seismic reflectors	Pyroclastic unit, uncertain in attribution, deposited from the offshore surrounding Capo Miseno to the Miseno volcanic bank, involved by wedging and growth, testifying its deposition during vertical downthrowing of normal faults.	
dk	Sub-vertical volcanic bodies, acoustically transparent and locally bounded by normal faults.	Volcanic dykes due to magma uprising in correspondence to normal faults.	Eastern and central Pozzuoli Gulf
pyr1	Discontinuous to parallel seismic reflectors.	Pyroclastic unit, uncertain in attribution, deposited in a structural depression under the volcanic edifice of Capo Miseno.	Eastern Pozzuoli Gulf
G3	Discontinuous to parallel seismic reflectors.	Lower sedimentary unit of the basin fill, composed of clastic deposits; strongly involved by wedging and growth in correspondence to anticlines (Punta Pennata anticline, Pozzuoli anticline, Nisida anticline) and synclines (central syncline of the Pozzuoli Gulf; Epitaffio syncline).	Pozzuoli Gulf
V3	Acoustically-transparent to discontinuous seismic unit; strongly eroded at its top.	Volcaniclastic unit related to the northern margin of the Pentapalummo bank; intensively deformed by anticlines (Punta Pennata anticline, Pozzuoli anticline, Nisida anticline) and synclines (central syncline of the Pozzuoli Gulf; Epitaffio syncline) individuated due to compressional deformation genetically related to main magmatic events.	Pozzuoli Gulf

In the geological interpretation of magnetic data the knowledge of magnetic rock properties requires an understanding of both magnetic susceptibility and remnant magnetization. The factors controlling rock magnetic properties for various types of rocks have been resumed by Reynolds et al. (1990) [132] and Clark and Emerson (1991) [43].

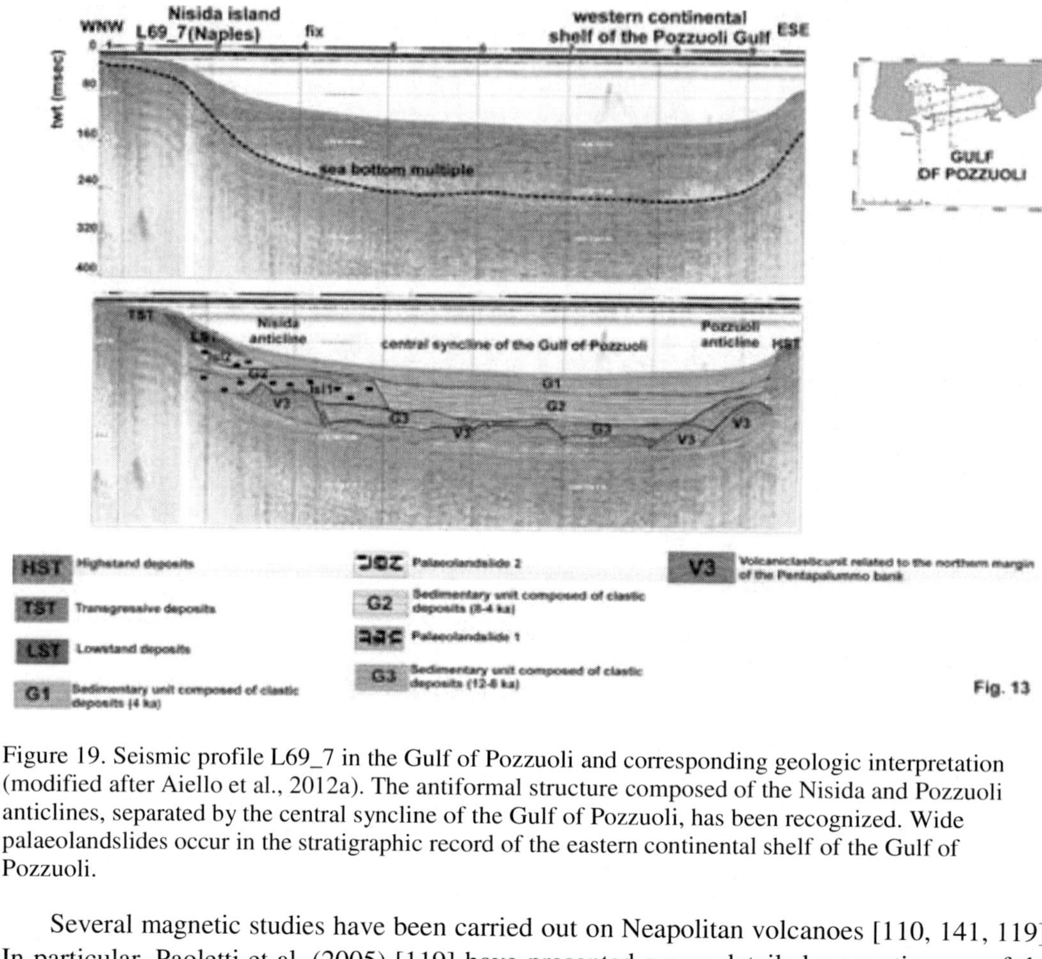

Figure 19. Seismic profile L69_7 in the Gulf of Pozzuoli and corresponding geologic interpretation (modified after Aiello et al., 2012a). The antiformal structure composed of the Nisida and Pozzuoli anticlines, separated by the central syncline of the Gulf of Pozzuoli, has been recognized. Wide palaeolandslides occur in the stratigraphic record of the eastern continental shelf of the Gulf of Pozzuoli.

Several magnetic studies have been carried out on Neapolitan volcanoes [110, 141, 119]. In particular, Paoletti et al. (2005) [119] have presented a new detailed magnetic map of the whole Neapolitan active volcanic district, obtained through the merging of airborne, land and marine magnetic data. The interpretation of magnetic dataset through magnetic analysis including inversion techniques has allowed for the characterization of buried volcanic structures [6], enabling a better understanding of the connection between tectonism and volcanism in the study area.

The magnetic survey of the Gulf of Pozzuoli has been recorded using an acquisition system allowing for the contemporaneous acquisition of magnetic data (proton magnetometer) and positioning (differential GPS) data hosted on a little boat [83]. During this period 994 positions of Earth Magnetic Field (EMF) have been collected in the Pozzuoli Bay (Figure 20). The survey in the Gulf of Pozzuoli has been realized through a proton magnetometer, G-856 model (Geometrics Inc.), having a precision of reading of 0.1 nT and an accuracy of 0.5 nT. The positioning of measurements was ensured by a short range differential system GPS-Motorola III. The measurements have been collected along N-S oriented tracks, having an average distance ranging between 200 and 300 m.

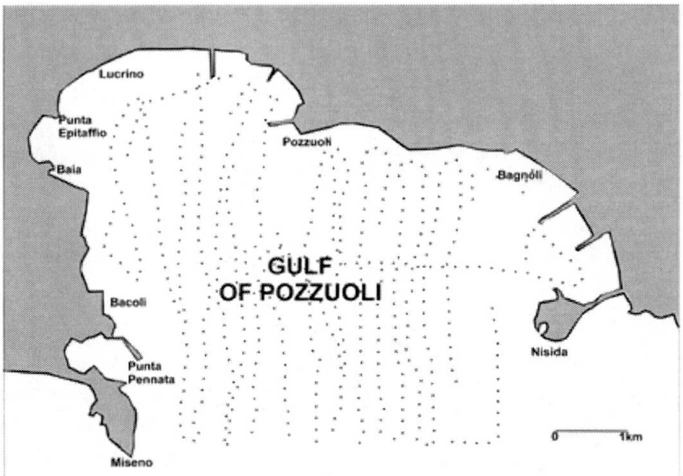

Figure 20. Location of the acquisition points of the Earth Magnetic Survey intensity in the Gulf of Pozzuoli (modified after Galdi et al., 1988; Aiello et al., 2012a).

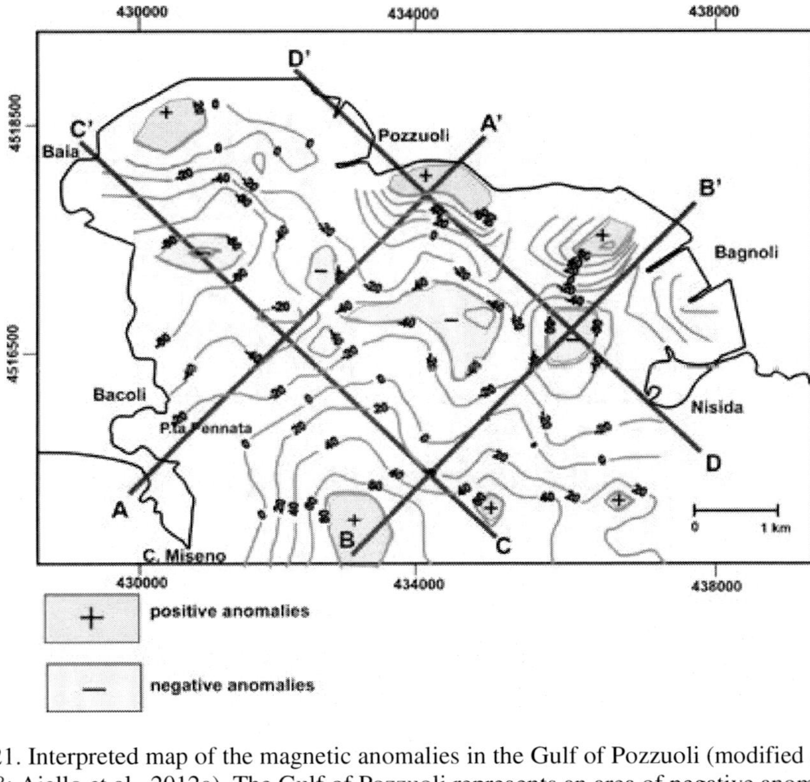

Fig. 19

Figure 21. Interpreted map of the magnetic anomalies in the Gulf of Pozzuoli (modified after Galdi et al., 1988; Aiello et al., 2012a). The Gulf of Pozzuoli represents an area of negative anomaly, excluding some areas of positive anomaly located off Baia, Pozzuoli and Bagnoli and some scattered positive anomalies on the continental shelf between Miseno and Nisida.

During the days of magnetic acquisition the check of magnetic data recorded by the Capri station and by L'Aquila Geomagnetic Observatory [96] has shown the lacking of magnetic perturbations having a short period. The data have not been corrected for the diurnal correction of the Earth Magnetic Field intensity. In fact, the measurements have been collected in a few hours during which the Earth Magnetic Field variations evidenced by the base station of Capri island were not significant.

An interpreted map of the magnetic anomalies in the Gulf of Pozzuoli has been constructed (Figure 21) [83, 10]. The interpretation of the magnetic anomaly map has evidenced areas characterized by positive anomalies (represented in yellow in the figure) and areas characterized by negative anomalies (represented in light yellow in the figure).

The inner continental shelf of the Gulf of Pozzuoli is covered by negative magnetic anomalies. The area surrounding the Pozzuoli harbour from the Caligola pier to the Pirelli jetty does not show significant magnetic anomalies. The area adjacent the Lucrino-Punta Pennata resort is characterized by a negative anomaly increasing southwards up to a magnetic minimum located near the Baia Castle (-100 nT; Figure 21).

The outer continental shelf of the Gulf of Pozzuoli is characterized by the alternance of magnetic maxima and minima (Figure 21). An area of magnetic maximum is located on a NE-SW oriented belt, about 1.7 km long (Figure 21). The inner shelf from Bagnoli to Pozzuoli shows two strong magnetic anomalies, separated by a thin belt having a normal magnetic value. Proceeding seawards two magnetic minima occur offshore the Bagnoli quarter; they are E-W trending and show values of – 40 and – 60 nT respectively. Another magnetic minimum has been observed in correspondence to the Baia Castle (-100 nT).

Four magnetic profiles, NE-SW and NW-SE oriented have been constructed (Figures. 22 and 23) [83, 10]. The magnetic profile A-A' ranges between Punta Pennata (Pozzuoli) to Via Napoli (town of Pozzuoli). It is characterized by a magnetic minimum of – 80 nT in the central area (corresponding to a water depth of – 90 m) and a magnetic maximum of 70 nT near the Pozzuoli shoreline.

The magnetic profile B-B' is shifted of 2.4 km towards SE with respect to the A-A'. Starting from SW it shows a monotonous magnetic trend since to the Nisida offshore (Naples), where a strong increase of gradient occurs. On the magnetic anomaly profile the magnetic maxima occurring proceeding landwards are not controlled from the geology, but from the occurrence of the industrial system of Bagnoli. It has controlled the release in the environment of metals derived by the burning of fossil coals and by the remnants of the industrial production. Another explanation may be suggested. The hydrothermal activity related to volcanism in the Phlegrean Fields could have controlled the supply in the environment of a large amount of metallic elements, causing the occurrence of the magnetic anomalies [26, 68].

The magnetic profile C-C'ranges between the Pentapalummo Bank, located on the outer continental shelf of the Gulf of Pozzuoli to Punta dell'Epitaffio. It is characterized by a strong magnetic anomaly followed by a related maximum near the Bagnoli quarter and by a magnetic mimimum near the Nisida island.

The magnetic profile D-D' ranges between the Nisida inlet and the historical centre of the Pozzuoli town (Rione Terra). A magnetic minimum has been recognized at 1.2 km from Nisida; by continuing, the value of intensity rapidly increases reaching the maximum value of 60 nT near the Rione Terra of Pozzuoli.

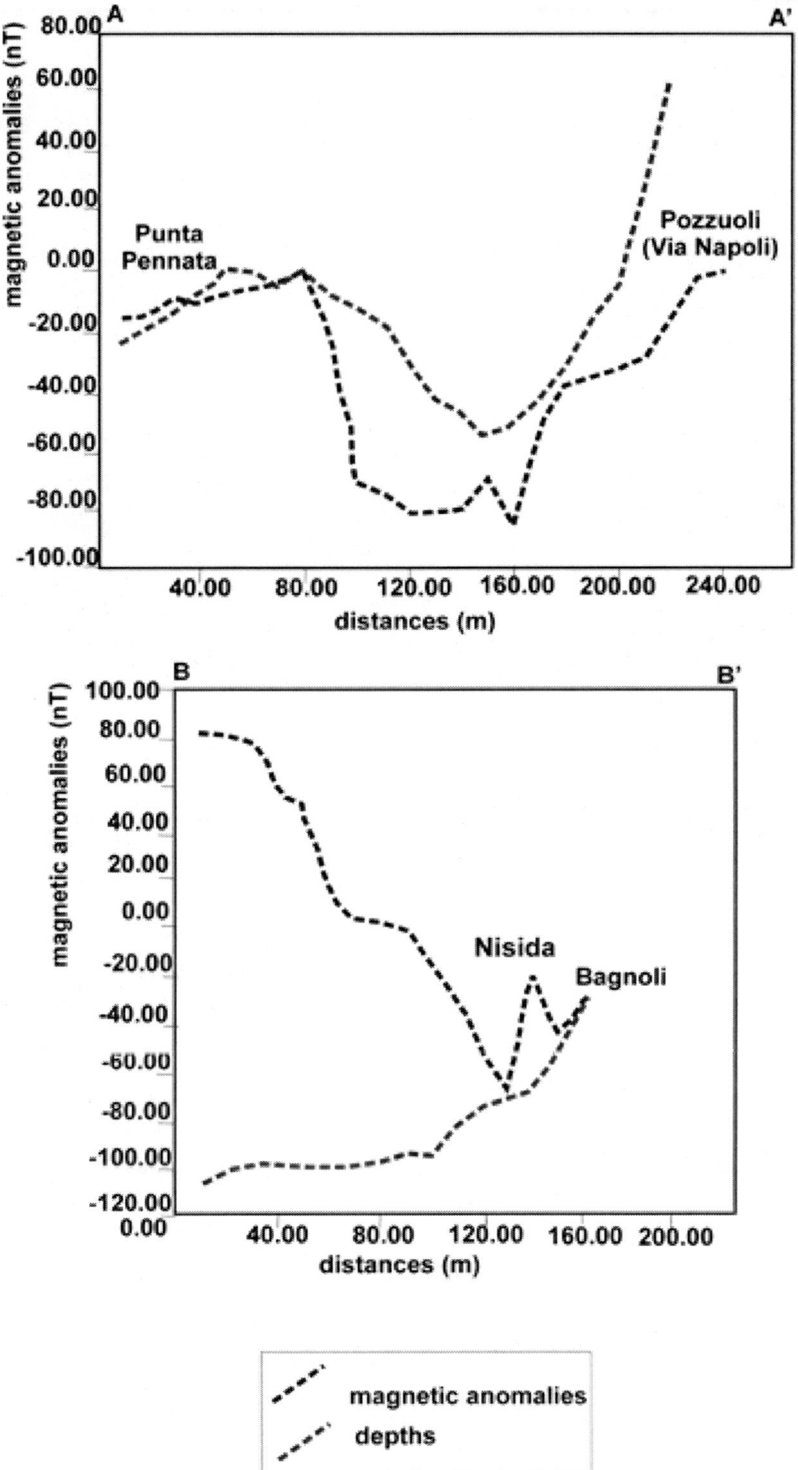

Figure 22. Magnetic sections of the iso-anomalies NE-SW oriented in the Gulf of Pozzuoli (modified after Galdi et al., 1988; Aiello et al., 2012a).

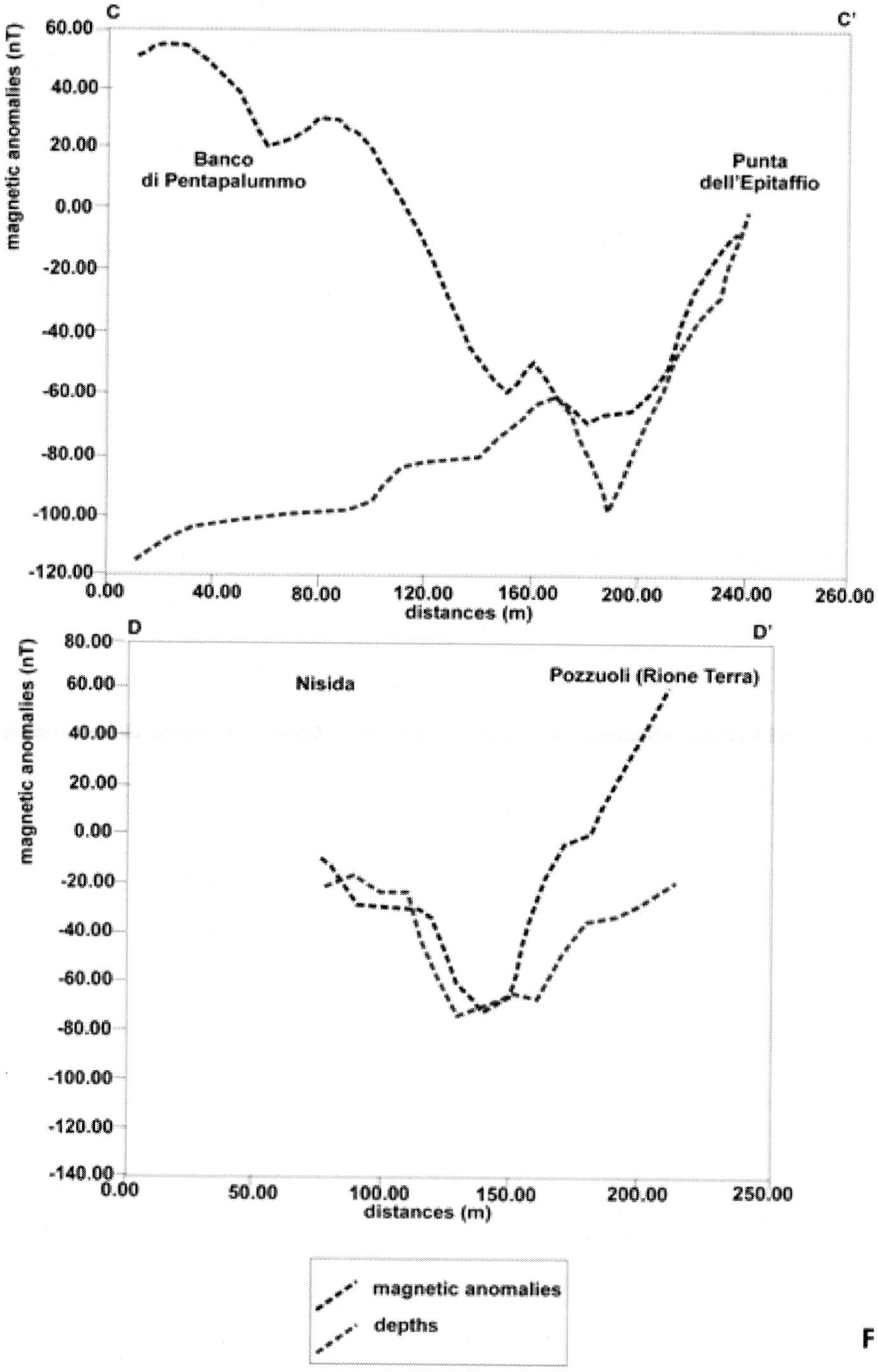

Figure 23. Magnetic sections of the iso-anomalies NW-SE oriented in the Gulf of Pozzuoli (modified after Galdi et al., 1988; Aiello et al., 2012a).

Offshore the Phlegrean Fields volcanic complex significant magnetic anomalies are located in a belt of submarine volcanic banks in the outer shelf of the Gulf of Pozzuoli (Figure 24). The interpreted maps shown in the figure evidence that the Phlegrean Fields offshore represents a relatively complex magnetic anomaly area, characterized by several magnetic anomaly fields having different intensities. Two dipolar anomalies categorized by a maximum-minimum couple have been identified. The first anomaly E-W oriented and located in the northern sector shows a minimum of – 200 nT related to a maximum of +185 nT. Such a values, not so high, are related with volcanic bodies not cropping out at the sea floor but buried by sediments. The second anomaly, NW-SE oriented and located in the eastern sector shows a maximum-minimum couple with a relative intensity similar to that of the previously discussed field. Other anomalies not dipolar and of lower intensity ranging between 40 and 135 nT are due to the occurrence of small volcanic edifices.

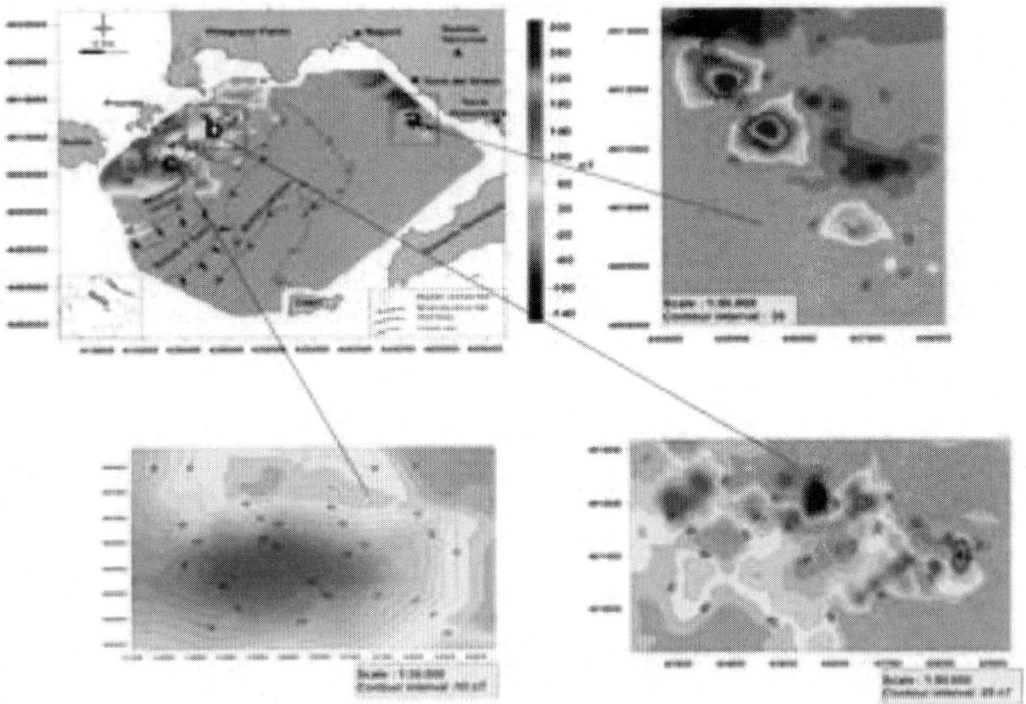

Figure 24. High resolution magnetic anomaly map of the Naples Bay (modified after Aiello et al., 2005; 2012a). The inset "a" represents a magnetic anomaly area located offshore the Somma-Vesuvius volcanic complex, while the inset "b" represents another magnetic anomaly area located offshore the Phlegrean Field volcanic complex. The inset "c" represents a magnetic anomaly area located on the continental slope of the Naples Bay, at the Magnaghi canyon's head.

In conclusion, the interpreted map of the magnetic anomalies in the Gulf of Pozzuoli has allowed us to distinguish both areas characterized by positive anomalies (represented in yellow) and areas characterized by negative anomalies (represented in light yellow).

The inner continental shelf of the Gulf of Pozzuoli is regarded as negative magnetic anomalies and the correlation with the volcanic structures evidenced by Sparker data is not clear. A more densely spaced magnetic survey needs to be recorded in order to identify the volcanic dykes shown by seismic profiles on marine magnetics. The volcaniclastic units identified on seismic profiles do not produce significant magnetic signatures probably due to their composition (tuffs rather than lavas).

5.3. Ischia

The Ischia island is located in the north-western sector of the Gulf of Naples and together with the Phlegrean Fields and the Somma-Vesuvius volcanic complex represents an active volcanic area in the Roman comagmatic province [85, 127, 70]. Next studies [114, 116, 117] have put the Ischia island in the Phlegrean Volcanic District, including also the Phlegrean Fields and the Ischia island. In the Ischia island the occurrence of thermal sources and the strong fumarolic and seismic activity have demonstrated that its magmatic system is still active.

The volcanic history of the Ischia island is strictly related to the extensional tectonic phases involving the western margin of the Apennines during the Plio-Quaternary, triggering the collapse of the Apenninic chain and the formation of the Tyrrhenian sea and the Campania Plain [87, 50]. The extension happened through normal faults; NW-SE trending and subordinately NE-SW trending separating the graben in blocks and allowing for the magma uprising [33, 80, 95, 78].

The Campania Plain is a graben having a NW-SE trending bounded by portions of Mesozoic carbonate platforms which during the Plio-Quaternary has undergone a subsidence of more than 5 kilometers. The graben is interrupted by buried structural highs having a ENE-WSW direction in correspondence to the Phlegrean caldera and to the Ischia and Procida islands. The volcanic activity controlled by fractures having a NE-SW direction and subordinately NW-SE direction has accompanied the extensional processes during the Plio-Quaternary. Regional tectonic processes have triggered the onset of recent volcanic activity along some preferential directions in the Ischia island [151]. In fact the fractures having a NE-SW direction characteristics of the Ischia-Procida-Phlegrean Fields elongment partly control the tectonic and eruptive history of the island.

The volcanic activity of the Ischia island has been marked by the event corresponding to the eruption of the Epomeo Green Tuffs (55 ky B.P.) [147, 128, 81, 151]. The morphological depression actually corresponding to the central part of the island represents a caldera related to the Epomeo Green Tuffs. This area filled by the Epomeo Green Tuffs and overlying marine deposits has been involved by phenomena of resurgence starting from 33 ky B.P. [115, 151, 1]. These processes have involved the only calderic center allowing for its downthrowing in blocks.

The modes of uplift and deformation of the Epomeo block have influenced the distribution of the active volcanic centers during the last 10 ky to the periphery of the block itself, being concentrated along its eastern margin [115, 1]. It is possible that the caldera resurgence was controlled by the intrusion of new magma in the volcanic system [114, 81, 115]. This intrusion should have determined an increase of the pressure in the upper part of the magmatic chamber and the emplacement of a stress field characterized by a main maximum strenghth oriented in a vertical direction and applied to the rocks at the top of the magmatic chamber.

Two models have been proposed to explain the mechanism of resurgence in the Ischia island [115, 105], representing a resurgent block rather than a resurgent dome. In the resurgent blocks, in fact, the volcanic and seismic activity develop mainly along the borders, as it happens for the Ischia island. The first proposed model [147, 81] is a model of volcano-tectonic horst, where the resurgent block is bordered by faults dipping outwards. The second one is a model based on a simple shear mechanism [115] with the resurgent block tilted around a horizontal axis having a N-S direction and bounded by high-angle faults dipping inwards (Figure 25) [115]. The second model describes the phenomenon of resurgence taking into account all the volcanologic and geologic constraints, such as the distribution of the eruptive centres linked to the resurgence. A similar mechanism determines the conditions for the magma uprising only in the sector of the resurgent area which is superimposed to an extensional stress regime observed also at Pantelleria and Phlegrean Fields. The lacking of eruptions in the northern sector during the last 33 ky B.P. suggests that the resurgence has locally substituted the volcanic activity. This observation is coherent with the fact that the mechanical energy requested for the resurgence has been evaluated with mirated structural models, equivalent to the thermal energy of a middle averaged eruption [1]. The simple shear model proposed by Orsi et al. (1991) [115] and rielaborated by Acocella and Funiciello (1999) [1] and Molin et al. (2003) [105] envisages that high-angle structures form at the beginning of the deformation with immersions towards the centre of the island, which define the margins of the resurgent block. These structures may represent the product of the reactivation of pre-existing faults. If the block is uplifted in absence of internal deformations void spaces will create. To contrast this possible instability it is necessary that the block is basculated rotating around a horizontal axis perpendicular to the direction of maximum strengthening and undergoing an inner deformation according to a simple shear mechanism.

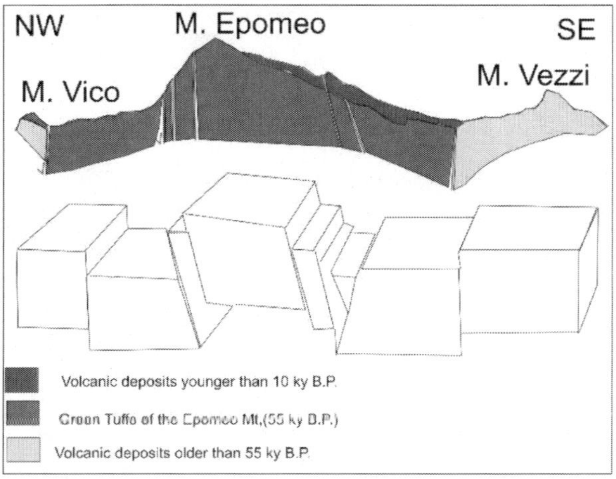

Figure 25. Model of the mechanism of resurgence of the Ischia island (modified after Orsi et al., 1991).

The maximum vertical deformation of the block is accommodated through the formation of high-angle reverse faults oriented parallel to the rotation axis allowing for a horizontal shortening and a compressional stress regime, preventing the magma uprising. Normal faults having the same orientation form along the opposite side of the block to accommodate the resulting shortening with an equivalent strengthening. This triggers the onset of an extensional stress regime making easier the magma uprising along this sector of the resurgent block.

The Ischia volcanic history and evolution of magmatic system are resumed. The beginning of the volcanic activity in the Ischia area is not precisely known, since the oldest dated rocks are not the oldest in outcrop. They have an age of about 150 ky B.P. [147] and represent the remnants of a volcanic edifice exposed in the south-eastern sector of the island.

They include a sequence of pyroclastic rocks interlayered by paleosoils and are buried by the products of next eruptions. Vezzoli (1988) [147] have distinguished five phases of activity in the volcanic history of Ischia island: 1) volcanism oldest than 150 ky B.P., representing by great pyroclastic eruptions; 2) volcanism ranging in age from 150 and 75 ky B.P., characterized by the activity of several lava domes and small monogenetic volcanic centres; 3) volcanism ranging in age between 55 and 33 ky B.P. The pyroclastic eruption of the Epomeo Green Tuffs, which has marked the eruptive history of Ischia island is dated back to 55 ky B.P. 4) volcanism ranging in age from 28 and 18 ky B.P. The activity is concentrated in the south-western sector of the island with scattered eruptive centres in the eastern sector and is characterized by monogenetic centres mainly effusive active during short time intervals. 5) Volcanism ranging in age between 10 ky B.P. and 1302 D.C. involving mainly the eastern sector of the island and including volcanic eruptions from many monogenetic centres.

Table 5. Seismic units in the northern Ischia offshore (Casamicciola)

Seismic unit	Seismic facies	Geological interpretation
V1	Acoustically transparent, wedge-shaped seismic unit	Volcanic unit of Casamicciola. Undetermined volcanic unit probably corresponding to a volcanic acoustic substratum, eroded at its top by a subaerial unconformity.
V2	Acoustically transparent, dome-shaped seismic unit	Volcanic domes of Casamicciola. Dome-shaped volcanic edifices, interstratified in the basin filling (C and D units)
C	Parallel to subparallel seismic reflectors and locally prograding clinoforms.	Lower seismic unit of the basin filling of the northern Ischia offshore.
D	Parallel to subparallel seismic reflectors.	Intermediate seismic unit of the basin filling of the northern Ischia offshore.
E	Parallel seismic reflectors with bidirectional onlaps.	Upper seismic unit of the basin filling of the northern Ischia offshore.
H1/H2	Hummocky seismic facies and wedge-shaped external geometry.	Deposits of volcanic gravity flows corresponding to two different events.
FST	Progradational seismic reflectors.	Forced regression prograding wedges.

Based on K/Ar dating Poli et al. (1987) [127] have individuated four phases of activity: 1) phase older than 150 ky B.P. characterized by pyroclastic products and interstratified lavas, without a clear relationship with a volcanic center. The occurrence of two paleosoils testifies prolonged periods of quiescence between the eruptions. 2) Phase ranging between 150 and 75 ky B.P. The extensional tectonics has determined the eruption of several lava domes in connection with the pyroclastic volcanism. The half-circle distribution of the domes seems to indicate the occurrence of the caldera centre. 3) Phase ranging between 55 and 20 ky B.P. This phase starts with the eruption of the Epomeo Green Tuffs, whose eruptive centre is located in the southern part of the island and continues in several eruptive centres. 4) Phase ranging between 10 ky B.P. and 1302 D.C. The activity is concentrated in the Ischia graben, coincident with the south-eastern sector. Poli et al. (1987) [127] have related to each phase the chemical characteristics of the products mainly classified as alkalitrachytes and subordinately as phonolites latites trachytes and trachybasalts. They have evidenced a correlation between the eruptive phases and the magmatic evolution of the island, defined by the trending of the trace elements during the geological time.

Table 6. Seismic units of the southern Ischia offshore (Ischia bank)

Seismic unit	Seismic facies	Geological interpretation
1	Progradational geometries with eroded topsets and preserved clinoforms.	Relict prograding wedge of Late Pleistocene marine deposits.
V3-CI	Acoustically transparent seismic facies; mounded-shaped external geometry.	Volcanic unit of the Ischia Channel. Pyroclastites and lavas genetically related to the relict volcanic edifices "I Ruommoli," "La Catena," "Le Formiche di Vivara" and "Il Pertuso" (eastern Ischia offshore).
3	Discontinuous seismic reflectors having a high amplitude.	Pyroclastic unit of uncertain attribution in lateral stratigraphic contact on the volcanic unit of the Ischia Channel next to "Il Pertuso" relict volcanic edifice.
V2-BI	Acoustically-transparent seismic facies, bank-shaped external geometry.	Volcanic unit of the Ischia Bank. Lavas and pyroclastic products genetically related to the monogenetic volcanic edifice of the Ischia Bank.
5	Discontinuous seismic reflectors having a high amplitude.	Pyroclastic products onlapping into depressions and channellised erosional morphologies genetically related to the last eruptive phases of the Procida island (Solchiaro Yellow Tuff).
FST/TST/HST	Prograding clinoforms (FST), retrogradational reflectors (TST), parallel seismic reflectors (HST).	Late Quaternary Depositional Sequence, organized as forced regression prograding wedges (FST) on the margins of the Ischia Bank, transgressive fillings of continental shelf (TST) overlain by highstand deposits (HST) downlapping on flooding surfaces.
7	Parallel seismic reflectors	Pelites and sandy pelites cropping out at the sea bottom in outer shelf domains.
Dx	Parallel seismic reflectors	Bioclastic sands and gravels cropping out at the sea bottom in correspondence to the Ischia Bank.

Civetta et al. (1991) [42] have distinguished three periods of activity during the last 55 ky. Each period was dominated by specific processes of differentiation and was characterized by the arrival of new magma less differentiated in the system. Three periods of activity are indicated by the different trend of trace element contents and by the Sr isotopic composition during geological time. 1) First period ranging between 55 and 33 ky B.P. starting with the ignimbritic eruption of the Epomeo Green Tuffs during which the caldera collapsed. The calderic depression flooded by the sea has been partly filled by the products of the eruption. The tuffs were partly deposited in a marine environment (Epomeo), partly in a subaerial environment (M. Vico, S. Angelo, Scarrupata di Barano). After this eruption the volcanism continued since 33 ky B.P. with magmatic and phreatomagmatic explosions, whose products crop out in the coastal cliffs of S.Angelo, Punta Imperatore, Citara and M.Vico. 2) Second period ranging between 28 and 18 ky B.P. It begins with the trachybasaltic eruption of Grotta di Terra along the south-eastern coast of the island and continues since 18 ky B.P. with the eruption of trachytic magmas. The products of explosive eruptions are mainly alkalitrachytes while the effusive eruptions have generated mainly trachytic lava fluxes. The rocks erupted during this period are exposed at Grotta del Mavone-Campotese M.te di Vezzi S. Anna Carta Romana Punta Imperatore and S. Angelo. The beginning of this period is marked by a significant difference among the chemical and isotopic compositions of the erupted magmas. The chemical composition varies during time by alkalibasaltic to alkalitrachytic with a contemporaneous increase of incompatible elements and of the Sr isotopic ratio. These variations are compatible with a model implying the arrival of new magma in the system its progressive differentiation and its mixing with the resident alkalitrachytic magma. The beginning of the third period of activity is fixed to 10 ky B.P. after a relatively long period of quiescence and continues since the last eruption in the 1302 D.C. (Arso lava flow; Chiesa et al., 1987) [38]. This period has been characterized by effusive eruptions which have emitted lava fluxes and domes and by magmatic and phreatomagmatic eruptions, which gave origin to pyroclastic fall deposits and small tuff cones. Most eruptive centres are located in the morphological depression eastwards to the Epomeo and include the volcanic edifices of Posta Lubrano Monte Rotaro Trippodi Costa Sparaina Montagnone Vateliero Molara and Cava Nocelle.

A grid of Sparker Multitip seismic profiles acquired in the Ischia offshore in the frame of research projects of submarine geological mapping has been interpreted adding new tectono-stratigraphic evidence on submarine prolongation of the Ischia volcanic complex [12].

Two sketch tables of the seismic units in the northern (Casamicciola) and southern Ischia offshore (Ischia Bank) have been constructed for this chapter (Tables 5 and 6).

Discussion and Conclusion

Marine geophysics of the Naples Bay has been discussed focusing on some *principles*, *applications* and *emerging technologies* in the use of seismic and marine magnetic data. The *applications* are represented by the Somma-Vesuvius offshore, by the Phlegrean Fields offshore and Gulf of Pozzuoli and by the Ischia volcanic complex.

The *principles* of seismic stratigraphy have been applied. As a general rule, the seismic stratigraphy is an analytical methodology allowing for the geological survey of the subsurface [146, 104]. This methodology has been developed from the end of the '70 of the last century and is particularly applied to the analysis of the continental margins. It is based on the checking of upper and lower terminations of seismic horizons and their depositional geometries (onlap, erosional truncation, downlap, toplap) with respect to the main unconformities, bounding specific intervals defined as depositional sequences. Type 1 and Type 2 sequence boundaries have been distinguished based on the interactions among eustatic rate and subsidence at the shelf margin. The seismo-stratigraphic approach makes possible the reconstruction of the tectono-stratigraphic evolution of a sedimentary basin related to eustatic oscillations.

In the application of the Naples Bay the occurrence of isolated volcanic bodies (intrusions domes volcanic necks and tabular seismic units acoustically transparent) makes the seismo-stratigraphic approach particularly complex in the geological interpretation of seismic profiles. While the volcanic bodies (lava flows domes intrusions) cannot be internally investigated through the reflection seismics because they are acoustically-transparent the seismic facies of the pyroclastic units may be observed due to their inner bedding. The marine deposition is sourced by alluvial and marine sediments and by volcanites and volcaniclastic deposits.

Some bathymetric maps have been constructed and are here presented (Figure 26). The data were recorded in 2001 onboard of the R/V Thetis (CNR) by using the Reson Seabat 8111 Multibeam sonar system, which properly works in the 0-1000 m depth range. The system interfaced with a Differential Global Positioning System was mounted on keel of Thetis R/V. It was composed of a ping source of 100 KHz 150° degree for the whole opening of the transmitted pulse and a 101 beams-receiver with a beam opening of 1.5°. Sound velocity profiles were regularly recorded and applied every 8 hours. Data were processed with the PDS2000 software /Reson-Thales) with a real-time acquisition control and partial beam exclusion filtering (particularly for what concern lateral beams) applied to data directly onboard based on the operators experience and the offline swath editing and de-spiking. The DTM generation and rendering of the whole dataset was subsequently reorganized in an MXN matrix (Digital Terrain Model, DTM) of 20x20 metres.

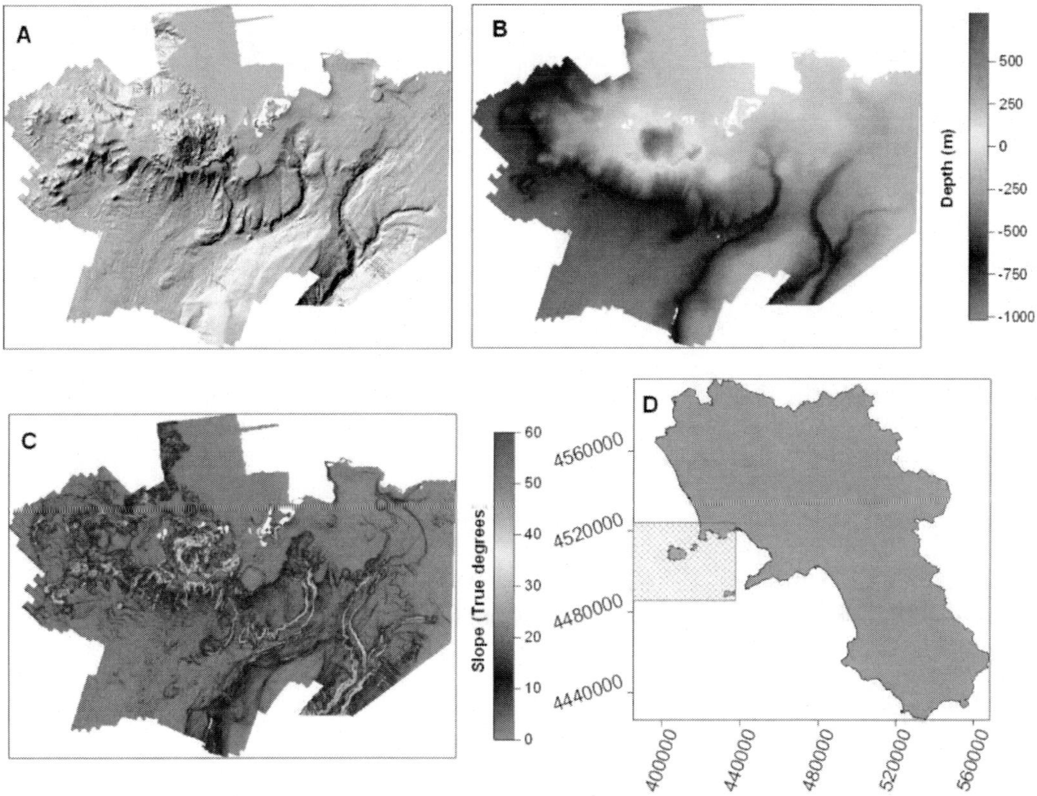

Figure 26. Bathymetric and slope maps of the Naples Bay.

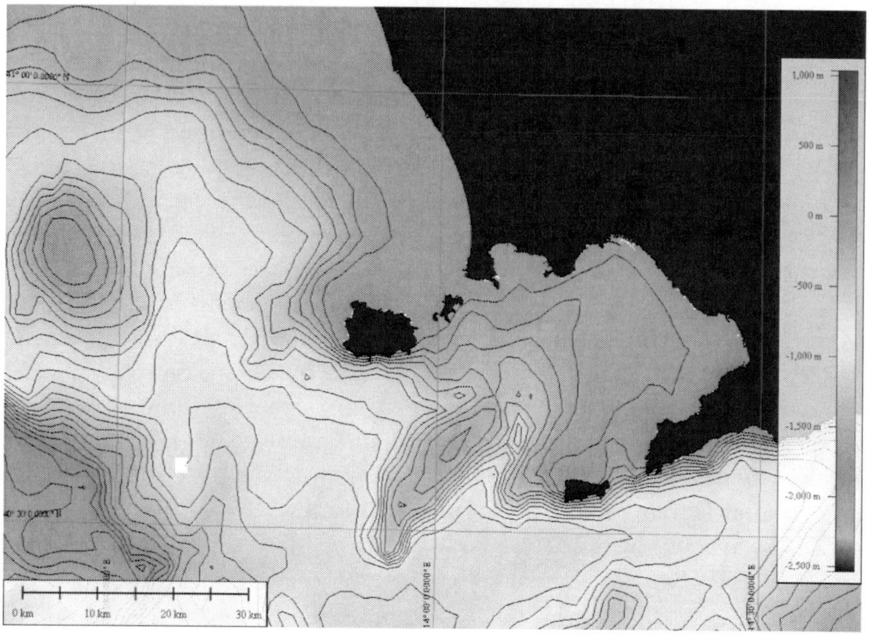

Figure 27. Sketch diagram showing the Ischia-Procida-Campi Flegrei volcanic complex, which is located over an important ENE-WSW regional structural alignment.

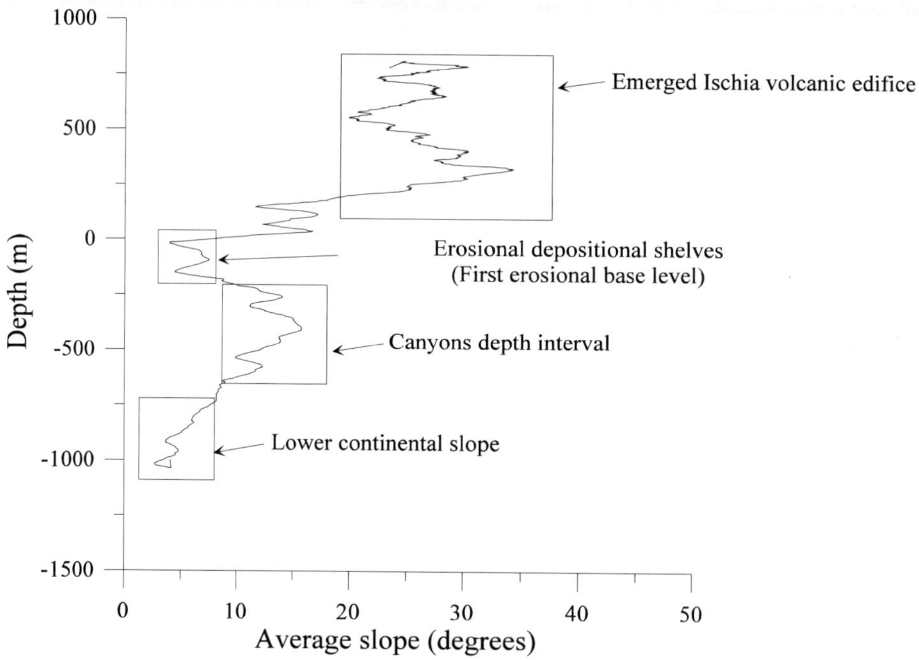

Figure 28. Sketch diagram showing elevation intervals versus average slopes offshore the Ischia island (modified after Aiello et al., 2012b).

Vertical accuracy is 1-5 cm for the used Reson Seabat 8111 Multibeam system (depending on the used acquisition range). The horizontal limit is a function of depth and of beam-number versus swath opening ratio, following the equation:

$$Nf = D*tg(SO/NB) \quad \text{(Equation 3)}$$

where: Nf = Nadir footprint,
D = depth,
SO = Swath Opening (in degrees),
NB = Number of beams in the used array.

The *application* of Ischia is discussed in detail. Large-scale bathymetric maps obtained by satellite data [14] have shown that the Phlegrean islands (Ischia Procida and Vivara) represent a sedimentologic boundary between the Gaeta Gulf to the north and the Naples Bay to the south. On the other side the Ischia-Procida-Campi Flegrei volcanic complex is located over an important ENE-WSW regional structural alignment which has controlled the emplacement of major eruptive centres of this volcanic sector also creating a morphological threshold between the Gaeta Gulf and the Naples Bay sedimentological units (Figure 27).

The elevation versus average slope plot [107] has been adopted to describe morphological macro-features of the Ischia offshore [11]. It has allowed to highlight where steep areas and flat areas occur in terms of elevation. The use of such a plot has been derived by the IAMC DTM and by slope map using an opportunely built routine (Figure 26).

The morphological macro-areas have been identified individuating groups of elevation/slope pairs that are attributed to specific settlements. The calculation has been carried out using 1 m depth window and then evaluating average slope value of all DTM cells that fallen inside each window (Figure 28). A median filter (25 points window) has been applied to smooth the progress of plot examination. In the Ischia offshore several morphological ranges have been identified. Each one is characterized by a well-defined elevation interval vs. average slope (Figure 28). They are listed above [11].

A) *The Ischia outcropping volcanic edifice* located at depth>0 and characterized by an average slope range of about 20°-40°.
   A2) *An intermediate stage* that acts as an according layer towards the continental shelf.
B) *The continental shelf* between the coastline and the 140-150 m (200 m in some cases) isobath (slopes 3°/8°).
C) *The upper continental slope* located between the platform edge and the 650 meters isobath (slopes 8°/20°).
D) *The lower continental slope* deeper than 650 meters in depth (average slopes 0°/8°).

These domains include several morphological elements correlating to a tectonic and/or a sedimentary process or to a volcanic event. In the platform as well as at greater depths terraces of abrasion and (or) deposition relict morphologies of volcanic edifices canyons and gullies can be recognized. The depositional shelf break is partially eroded at the head of some canyons (Cuma and Punta Cornacchia canyons). On the contrary to what recorded in normal depth distributions the increasing of dip angles in the lower portion of the A physiographic/geomorphologic unit has been outlined probably due to base faulting of Epomeo resurgent block. Drainage patterns particularly as regard to submarine canyons are present on A and C macro-units acting as morphological/sedimentological links between different ranges.

Slope instability on the flank of volcanoes is a well known phenomenon. It is magnified by the inner nature of the volcano buildings and is mainly due to their mechanism of emplacement to the hydrothermal vents (when present) to very fast accretion of edifices and to the different lateral erosional rate due to the different nature of products on the slopes [143]. The Ischia slope instability has been investigated through several studies highlighting the existence of catastrophic landslides (debris avalanches) particularly in the southern sector (the Ischia Debris Avalanche, IDA) [39, 53]. Here a giant (1.5 km^3) catastrophic event with a southwards dispersal axis occurred in the last 5 ky B.P. Other catastrophic events were detected both on the western [29] and the northern [53] sea sectors of the island. Such an instability could be triggered by the oversteepening generated by the uplift of the Epomeo block [65]. It seems significant that areas characterized by the presence of hummocky facies in the western sector of Ischia are different from those located in the north and south for the lack of consistency with areas of mass gap (border of detachment).

The Ischia Island is encircled by several lava outcrops interpreted as megadykes whose formation seems to be linked to the calderic nature of the volcanic complex (e.g., Punta Cornacchia, Zaro lava flow, Monte S. Angelo, Punta del Soccorso) [133]. Debris avalanche detachments are systematically developed between these volcanic features both in the southern and in the northern sides of the island but not in the western one. In fact lateral collapses that characterize this area seem to be originated within morphological protrusions (lava outcrops). In addition it is remarkable that debris deposits are systematically originated inside the A and C morphological units whereas they are lacking within the C and D units. On the contrary slow mass movement at the border of detachments have been detected only into the D unit area.

This evidence has been interpreted as it follows:

A is *mainly related to the rise of the Epomeo block;*

B is mainly related to the occurrence of the upper shelf, strongly eroded on its upper border;

C individuates the upper scarp, which is tectonically-controlled, since it exhibits an average slope value higher than the expected one and a strong gradient at his foot;

D outlines the lower escarpment, which is characterized by the presence of a hemipelagic sedimentary cover.

The seismic exploration of the Naples Bay has been performed mainly through Sparker systems and Watergun sources. The evolution of seismic source capability in terms of technological advances together with the improvement of techniques of data processing has permitted the development of high resolution stratigraphic studies on deep and intermediate geological setting of the Naples Bay. Among the most important historical results there is the seismic survey of the Naples and Pozzuoli Bay recorded through Sparker EG&G (8 kJ) and Boomer systems [91, 21]. The evolution towards new technologies has allowed for the application of Multitip Sparker, Airgun and Watergun seismic sources which as described in this chapter have allowed for a detailed tectono-stratigraphic reconstruction of Somma-Vesuvius offshore Phlegrean Fields offshore Gulf of Pozzuoli and Ischia offshore.

REFERENCES

[1] Acocella, V. & Funiciello, R. (1999). The interaction between regional and local tectonics during resurgent doming: the case of the island of Ischia, Italy. *Journal of Volcanology and Geothermal Research*, *10* (2), 109-123.

[2] Agip, (1977). Temperature sotterranee. Inventario dei dati raccolti dall'AGIP durante la ricerca e la produzione di idrocarburi in Italia. Ed. AGIP, pp. 1930.

[3] Agip, (1981). Italia, Carta Magnetica – Anomalie del campo magnetico residuo, scala 1:500.000. S. Donato Milanese, Italy.

[4] Aiello, G. & Marsella, E. (2014). The Southern Ischia canyon system: examples of deep sea depositional systems on the continental slope off Campania (Italy). *Rendiconti online della Società Geologica Italiana*, doi 10.3301/ROL.2014.145.

[5] Aiello, G., Angelino, A., Marsella, E., Ruggieri, S. & Siniscalchi, A. (2004). Carta magnetica di alta risoluzione del Golfo di Napoli. *Bollettino della Società Geologica Italiana*, *123*, 333-342.

[6] Aiello, G., Angelino, A., D'Argenio, B., Marsella, E., Pelosi, N., Ruggieri, S. & Siniscalchi, A. (2005). Buried volcanic structures in the Gulf of Naples (Southern Tyrrhenian sea, Italy) resulting from high resolution magnetic survey and seismic profiling. *Annals of Geophysics*, *48* (6), 1-15.

[7] Aiello, G., Marsella, E. & Ruggieri, S. (2010). Three-dimensional magneto-seismic reconstruction of the "Torre del Greco" submerged volcanic structure (Naples Bay, Southern Tyrrhenian sea, Italy): implications for Vesuvius' marine geophysics and volcanology. *Near Surface Geophysics*, *8* (1), 17-31.

[8] Aiello, G., Cicchella, A. G., Di Fiore, V. & Marsella, E. (2011a). New seismo-stratigraphic data of the Volturno Basin (northern Campania, Tyrrhenian margin, southern Italy): implications for tectono-stratigraphy of the Campania and Latium sedimentary basins. *Annals of Geophysics*, *54* (3), 265-283.

[9] Aiello, G., Marsella, E., Cicchella, A. G. & Di Fiore, V. (2011b). New insights on morpho-structures and seismic stratigraphy of the Campania continental margin based on deep multichannel profiles. *Rendiconti Lincei Sc. Fis. e Nat.*, *22* (4), 349-373.

[10] Aiello, G., Marsella, E. & Di Fiore, V. (2012a). New seismo-stratigraphic and marine magnetic data of the Gulf of Pozzuoli (Naples Bay, Tyrrhenian sea, Italy): inferences for the tectonic and magmatic events of the Phlegrean Fields volcanic complex (Campania). *Marine Geophysical Researches*, *33* (2), 93-125.

[11] Aiello, G., Giordano, L., Marsella, E. & Passaro, S. (2012b). Seismic stratigraphy and marine magnetics of the Naples Bay (Southern Tyrrhenian sea, Italy): the onset of new technologies in marine data acquisition, processing and interpretation. In: *Stratigraphic Analysis of Layered Deposits* (Elitok O. Ed.), Chapter 2, pp. 21-60, Intech Science Publishers, Croatia.

[12] Aiello, G., Marsella, E. & Passaro, S. (2012c). Stratigraphic and structural setting of the Ischia volcanic complex (Naples Bay, Southern Italy) revealed by submarine seismic reflection data. *Rendiconti Lincei*, *23*, 387-408.

[13] Alessio, G., De Falco, M., Di Crescenzo, G., Nappi, R. & Santo, A. (2013). Flood hazard of the Somma-Vesuvius region based on historical (19-20th century) and geomorphological data. *Annals of Geophysics*, *56* (4), S0434, doi: 10.4401/ag-6440.

[14] Amante, C. & Eakins, B. W. (2008). ETOPO1 Arc-Minute Global Relief Model: Procedures, Data Sources and Analysis. National Geophysical Data Center, NESDIS, NOOA, U.S. Department of Commerce, Boulder, CO.

[15] Amoruso, A. & Crescentini, L. (2013). Analytical models of volcanic ellipsoidal expansion sources. *Annals of Geophysics*, *56* (4), S0435, doi 10.4401/ag-6441.

[16] Arnò, V., Principe, C., Rosi, M., Santacroce, R. & Sheridan, M. F. (1987). Eruptive history. In: *Somma-Vesuvius*, (Santacroce R. Ed.), CNR, Quaderni De La Ricerca Scientifica, Roma, Italy.

[17] Auger, M., Gasparini, P., Virieux, J. & Zollo, A. (2001). Seismic evidence of an extended magmatic sill under Mt. Vesuvius. *Science*, *294*, 1510-1512.

[18] Barberi, F., Cassano, S., La Torre, P. & Sbrana, A. (1991). Structural evolution of Campi Flegrei caldera in light of volcanological and geophysical data. *Journal of Volcanology and Geothermal Research*, *48* (1/2), 33-49.

[19] Belkin, H. E., De Vivo, B., Roedder, E. & Cortini, M. (1985). Fluid inclusion geobarometry from ejected Mt. Somma-Vesuvius nodules. *American Mineralogy*, *70*, 288-303.

[20] Belkin, H. E. & De Vivo, B. (1993). Fluid inclusion studies of ejected nodules from plinian eruptions of Mt. Somma-Vesuvius. *Journal of Volcanology and Geothermal Research*, *58*, 98-100.

[21] Bernabini, M., Latmiral, G., Mirabile, L. & Segre, A. G. (1971). Alcune prospezioni sismiche per riflessione nei Golfi di Napoli e Pozzuoli. *Rapp. Comm. Int. Mer. Medit.*, *21*, 929-934.

[22] Berrino, G., D'Errico, V. & Ricciardi, G. (2013). Thirty years of precise gravity measurements at Mt. Vesuvius: an approach to detect underground mass movements. *Annals of Geophysics*, *56* (4), SO436, doi:10.4401/ag-6442.

[23] Biagio, G. (2007). Studio vulcanologico delle porzioni sommerse del Distretto Vulcanico Flegreo attraverso indagini di sismica marina ad alta risoluzione e correlazione delle facies alle interfacce terra-mare. *PhD Thesis Scienze della Terra*, University of Pisa, 126 pp.

[24] Bigi, G., Cosentino, D., Parotto, M., Sartori, R. & Scandone, P. (1992). Structural Model of Italy. *Monografie Progetto Finalizzato Geodinamica*, CNR, Roma, Italy.

[25] Blockley, S. P. E., Ramsey, C. B. & Pyle, D. M. (2008). Improved age modeling and high precision age estimate of late Quaternary tephras for accurate palaeoclimate reconstruction. *Journal of Volcanology and Geothermal Research*, *177*, 251-262.

[26] Bodnar, R. J., Cannatelli, C., De Vivo, B., Lima, A., Belkin, H. E. & Milia, A. (2007). Quantitative model for magma degassing and ground deformation (bradyseism) at Campi Flegrei, Italy: implications for future eruptions. *Geology*, *35* (9), 791-797.

[27] Brocchini, F., Principe, C., Castradori, D., Laurenzi, M. A. & Goria, L. (2001). Quaternary evolution of the southern sector of the Campanian Plain and early Somma-Vesuvius activity: insights from the Trecase well. *Mineralogy and Petrology*, *73*, 67-91.

[28] Bruno, P. P. G. & Rapolla, A. (1999). Study of sub-surface structure of Somma-Vesuvius (Italy) by seismic reflection data: implications for Campanian volcanism. *Tectonophysics*, *372*, 193-213.

[29] Budillon, F., de Alteriis, G., Ferraro, L., Marsella, E., Molisso, F., Tonielli, R. & Violante, C. (2003). I fondali dell'isola d'Ischia: morfobatimetria e geologia tra Punta del Chiarito e Punta San Pancrazio. In: Gambi M.C., De Lauro M., Iannuzzi F. (Eds.) Ambiente Marino Costiero e Territorio delle Isole flegree (Ischia, Procida e Vivara – Golfo di Napoli): risultati di uno studio multidisciplinare. *Memorie Acc. Sc. Fis. e Mat. in Napoli*, *5*, 83-102.

[30] Capuano, P., Russo, G. & Scarpa, R. (2013). P-wave velocity and density structure beneath Mt. Vesuvius: a magma body in the upper edifice. *Annals of Geophysics*, *56* (4), S0437, doi:10.4401/ag-6443.

[31] Caratori Tontini, F., Faggioni, O. & Giori, I., Stefanelli, P. (2003). La nuova carta aeromagnetica d'Italia. Confronto con i principali lineamenti tettonico-strutturali *Atti 22° Convegno Nazionale GNGTS*, 19-21 Novembre 2003.

[32] Carlino, S., Cubellis, E., Luongo, G. & Obrizzo, F. (2006). On the mechanics of caldera resurgence of Ischia island (southern Italy). In: *Mechanisms of activity and unrest of large calderas* (Troise C., De Natale G., Kilburn C, Eds.), Geological Society of London Spec. Publ., 269, 181-193.

[33] Carrara, E., Iacobucci, F., Pinna, E. & Rapolla, A. (1973). Gravity and magnetic survey of the Campanian volcanic area, Southern Italy. *Bollettino di Geofisica Teorica Applicata*, 15 (57), 39-51.

[34] Cassano, E. & La Torre, P. (1987). Geophysics. In: *Somma-Vesuvius* (Santacroce R., Ed.), CNR, Quaderni de la Ricerca Scientifica, Roma, Italy.

[35] Castellano, M., Buonocunto, C., Capello, M. & La Rocca, M. (2002). Seismic surveillance of active volcanoes: the Osservatorio Vesuviano seismic network (OVSN-Southern Italy). *Seismology Research Letters*, 73, 177-184.

[36] Chiappini, M., Meloni, A., Boschi, M., Faggioni, O., Beverini, N., Carmisciano, C. & Marson, I. (2000). Shaded relief magnetic anomaly map of Italy and surrounding marine areas. *Annals of Geophysics*, *43*, 983-990.

[37] Chiesa, S., Cornette, Y., Forcella, F., Gillot, P. Y., Pasquarè, G. & Vezzoli, L. (1985). Carta Geologica dell'Isola d'Ischia. *Monografie Progetto Finalizzato Geodinamica*, CNR, Roma, Italy.

[38] Chiesa, S., Civetta, L., De Lucia, M., Orsi, G. & Poli, S. (1987). Volcanological evolution of the island of Ischia. In: *The volcanoclastic rocks of Campania (southern Italy)*, (De Girolamo P., Ed.) Rendiconti Acc. Sc. Fis. e Mat. In Napoli Special Issue, pp. 69-83.

[39] Chiocci, F. L. & de Alteriis, G. (2006). The Ischia debris avalanche: first clear submarine evidence in the Mediterranean of a volcanic island prehistorical collapse. *Terra Nova*, *18*, 202-209.

[40] Cioni, R., Bertagnini, A., Santacroce, R. & Andronico, D. (2009). Explosive activity and eruption scenarios at Somma-Vesuvius (Italy): Towards a new classification scheme. *Journal of Volcanology and Geothermal Research*, *178*, 331-346.

[41] Cioni, R., D'Oriano, C., Bertagnini, A. & Andronico, D. (2013). The 2nd and 4th century explosive activity of Vesuvius: new data on the timing of the upward migration

of the post A.D. 79 magma chamber. *Annals of Geophysics*, *56* (4), S0438, doi: 10.4401/ag6444.

[42] Civetta, L., Gallo, G. & Orsi, G. (1991). Sr and Nd isotope and trace element constraints on the chemical evolution of the magmatic system of Ischia (Italy) in the last 55 ky. *Journal of Volcanology and Geothermal Research*, *46*, 213-230.

[43] Clark, D. A. & Emerson, D. W. (1991). Notes on rock magnetization characteristics in applied geophysical studies. *Exploration Geophysics*, *22*, 547-555.

[44] Cole, R. H. (1965). Underwater Explosions. Dover Publications, New York.

[45] Corradi, N., Ferrari, M., Giordano, F., Giordano, R., Ivaldi, R. & Sbrana, A. (2009). SAM source and D-seismic system: the Use in Marine Geological Mapping C.A.R.G. and P.N.R.A. Projects. *27th IAS Meeting of Sedimentologists*, Alghero (Italy), pp. 85-90, Extended Abstract.

[46] Coticchia, A., De Santis, A., Di Ponzio, A., Dominici, G., Meloni, A., Pierozzi, M. & Sperti, M. (2001). La Rete Magnetica Italiana e la Carta Magnetica d'Italia al 20.000. *Estratto dal "Bollettino di Geodesia e Scienze Affini,"* Rivista dell'Istituto Geografico Militare, anno LX, n. *4*.

[47] Cubellis, E., Ferri, M., Luongo, G. & Obrizzo, F. (2001). The roots of Mt. Vesuvius from gravity anomalies. *Mineralogy and Petrology*, *73*, 23-38.

[48] Cubellis, E. & Marturano, A. (2013). Felt index, source parameters and ground motion evaluation for earthquakes at Mt. Vesuvius. *Annals of Geophysics*, *56* (4), S0439, doi: 10.4401/ag-6445.

[49] Cusano, P., Petrosino, S., Bianco, F. & Del Pezzo, E. (2013). The first long period earthquake detected in the background seismicity at Mt. Vesuvius. Annals of Geophysics, 56 (4), S0440,doi: 10.4401/ag 6447.

[50] D'Argenio, B., Pescatore, T. S. & Scandone, P. (1973). Schema geologico-strutturale dell'Appennino meridionale (Campania e Lucania). *Quaderni dell'Accademia Nazionale dei Lincei, Problemi Attuali di Scienza e Cultura, 183*, 49-72.

[51] D'Argenio, B., Aiello, G., de Alteriis, G., Milia, A., Sacchi, M., Tonielli, R., Angelino, A., Budillon, F., Chiocci, F. L., Conforti, A., De Lauro, M., De Martino, G., D'Isanto, C., Esposito, E., Ferraro, L., Innangi, S., Insinga, D., Iorio, M., Marsella, E., Molisso, F., Morra, V., Passaro, S., Pelosi, N., Porfido, S., Raspini, A., Ruggieri, S., Sarnacchiaro, G., Terranova, C., Vilardo, G. & Violante, C. (2004). Digital Elevation Model of the Naples Bay and adjacent area (Eastern Tyrrhenian Sea, Italy). *Atlante di Cartografia Geologica "Mapping Geology in Italy,"* APAT, Rome, Italy.

[52] D'Auria, L., Esposito, A. M., Lo Bascio, D., Ricciolino, P., Giudicepietro, F., Martini, M., Caputo, T., De Cesare, W., Orazi, M., Peluso, R., Scarpato, G., Buonocunto, C., Capello, M. & Caputo, A. (2013). The recent seismicity of Mt. Vesuvius: inference on seismogenic processes. *Annals of Geophysics*, *56* (4), S00442, doi:10.4401/ag.6448.

[53] De Alteriis, G. & Violante, C. (2009). Catastrophic landslides off Ischia volcanic island (Italy) during prehistory. In: *Geohazard in Rocky Coastal Areas*, (Violante C., Ed.), Geological Society of London Spec. Publ., *322*, 73-104.

[54] De Bonatatibus, A., Latmiral, G., Mirabile, L., Palumbo, A., Sarpi, E. & Scalera, A. (1970). Rilievi sismici per riflessione: strutturali, ecografici (fumarole) e batimetrici del Golfo di Pozzuoli. *Bollettino Società Naturalisti in Napoli*, *79*, 97-115.

[55] Del Pezzo, E. & Bianco, F. (2013). Inside Mt. Vesuvius: a new method to look at seismic (velocity and attenuation) tomographic imaging. *Annals of Geophysics*, *56* (4), S0443, doi:10.4401/ag. 6449.

[56] Del Pezzo, E., Bianco, F., Castellano, M., Cusano, P., Galluzzo, D., La Rocca, M. & Petrosino, S. (2013). Detection of seismic signals from background noise in the area of Campi Flegrei: limits of the present seismic monitoring. *Seismological Research Letters*, *84*, 190-198.

[57] Deino, A. L., Orsi, G., De Vita, S. & Piochi, M. (2004). The age of Neapolitan Yellow Tuff caldera forming eruption (Campi Flegrei caldera, Italy) assessed by ^{40}Ar/^{39} Ar dating method. *Journal of Volcanology and Geothermal Research*, *133*, 157-170.

[58] Dello Iacono, D., Zollo, A., Vassallo, M., Vanorio, T. & Judenherc, S. (2009). Seismic images and rock properties of the very shallow structure of Campi Flegrei caldera (southern Italy). *Bulletin of Volcanology*, *71*, 275-284.

[59] De Natale, G., Troise, C., Pingue, F., De Gori, P. & Chiarabba, C. (2001). Structure and dynamics of the Somma-Vesuvius volcanic complex. *Mineralogy and Petrology*, *73* (1-3), 5-22.

[60] De Natale, G., Troise, C., Trigila, R., Dolfi, D. & Chiarabba, C. (2004). Seismicity and 3D substructure at Somma-Vesuvius volcano: evidence for magma quencing. *Earth and Planetary Science Letters*, *221*, 181-196.

[61] De Natale, P., De Natale, G., Gagliadi, G., Gianfrani, L. & Rocco, A. (2005). Novel laser-based techniques for monitoring of volcanoes. *Annals of Geophysics*, *48*, 205-221.

[62] De Natale, G. (2007). *Corso su 'Volcanic risk and eruption prediction'*, alla 'Conference and School on Predictability of Natural Disasters for our Placet in Ranger. A System View: Theory, Modes, data Analysis', Int. Center for Theoretical Physics 'Abdus Salam', Miramare, Trieste, 25 June-7 July 2007.

[63] De Pippo, T., Di Cara, A., Guida, M., Pescatore, T. & Renda, P. (1984). Contributi allo studio del Golfo di Pozzuoli. Lineamenti di geomorfologia. *Memorie della Società Geologica Italiana*, 27, 151-159.

[64] De Vita, S., Esposito, B. & Mirabile, L. (1979). Criteri di Progetto di Sparker a cortina per sismica ad alta risoluzione. *Atti del convegno Scientifico Nazionale Progetto Finalizzato Oceanografia e Fondi Marini.*

[65] De Vita, S., Sansivero, F., Orsi, G. & Marotta, E. (2006). Cyclical slope instability and volcanism related to volcano-tectonism in resurgent calderas: The Ischia island (Italy) case study. *Engineering Geology*, *86*, 148-165.

[66] De Astis, G., Pappalardo, L. & Piochi, M. (2004). Procida volcanic history: new insights into the evolution of the Phlegrean Volcanic District (Campania, Italy). *Bulletin of Volcanology*, *66*, 622-641.

[67] De Vivo, B., Rolandi, G., Gans, P. B., Calvert, A., Bohrson, W. A., Spera, F. J. & Belkin, H. E. (2001). New constraints on the pyroclastic eruptive history of the Campanian volcanic plain (Italy). *Mineralogy and Petrology*, *73*, 47-65.

[68] De Vivo, B., Belkin, H. E. & Lima, A. (2008). Environmental geochemistry: site characterization, data analysis and case histories. Elsevier, Amsterdam, 21 July 2008.

[69] Di Girolamo, P., Ghiara, M. R., Lirer, L., Munno, R., Rolandi, G. & Stanzione, D. (1984). Vulcanologia e petrologia dei Campi Flegrei. *Bollettino della Società Geologica Italiana*, *103*, 349-413.

[70] Di Girolamo, P., Melluso, L., Morra, V. & Secchi, F. (1985). Evidence of interaction between mafic and differentiated magmas in the youngest phase of activity at Ischia island (Italy). *Periodico di Mineralogia*, *64*, 393-411.

[71] Di Maio, R., Cecere, G., De Martino, P. & Piegari, E. (2013). Electric effects induced by artificial seismic sources at Somma-Vesuvius volcano. *Annals of Geophysics*, *56* (4), S0445, doi:10.4401/ag-6541.

[72] Di Vito, M. A., Isaia, R., Orsi, G., Southon, J., De Vita, S., D'Antonio, M., Pappalardo, L. & Piochi, M. (1999). Volcanism and deformation since 12.000 years at the Campi Flegrei caldera (Italy). *Journal and Volcanology and Geothermal Research*, *91*, 221-246.

[73] Doglioni, C. (1991). A proposal of kinematic modelling for W-dipping subductions – Possible applications to the Tyrrhenian-Apennines system. *Terra Nova*, *3*, 423-434.

[74] Dvorak, J. J. & Mastrolorenzo, G. (1991). The mechanisms of recent vertical crustal movements in Campi Flegrei caldera (Italy). *Journal of Volcanology and Geothermal Research, 91*, 221-246.

[75] Edgerton, H. E. & Hayward, G. G. (1964). The boomer sonar source for seismic profiling. *Journal of Geophysical Research*, *68*, 3033-3042.

[76] Esposito, A., D'Auria, L., Giudicepietro, F., Caputo, T. & Martini, M. (2013). Neural analysis of seismic data: applications to the monitoring of Mt. Vesuvius. *Annals of Geophysics*, *56* (4), S0446, doi:10.4401/ag6452.

[77] Esposti Ongaro, T., Neri, A., Todesco, M. & Macedonio, G. (2002). Pyroclastic flow hazard at Vesuvius from numerical modelling II. Analysis of local flow variables. *Bulletin of Volcanology*, *64*, 178-191.

[78] Faccenna, C., Funiciello, R. & Mattei, M. (1994). Late Pleistocene N-S shear zones along the Latium-Tyrrhenian margin: structural characteristics and volcanological implications. *Bollettino di Geofisica Teorica Applicata*, *36*, 507-522.

[79] Federico, C., Madonia, P., Cusano, P. & Petrosino, S. (2013). Groundwater geochemistry of the Mt. Vesuvius area: implications for seismic surveillance and relationship with hydrological and seismic signals. *Annals of Geophysics*, *56* (4), S0447, doi:10.4401/ag-6453.

[80] Finetti, I. & Morelli, C. (1973). Esplorazione sismica per riflessione nei Golfi di Napoli e Pozzuoli. *Bollettino di Geofisica Teorica Applicata*, *16*, 175-222.

[81] Fusi, N., Tibaldi, A. & Vezzoli, L. (1990). Vulcanismo, risorgenza calderica e relazioni con la tettonica regionale nell'isola d'Ischia. *Memorie della Società Geologica Italiana*, *45*, 971-980.

[82] Fusi, N., Mirabile, L., Camerlenghi, A. & Ranieri, G. (1991). Marine geophysical survey of the Gulf of Naples (Italy): relationships between submarine volcanic activity and sedimentation. *Memorie della Società Geologica Italiana*, *47*, 95-114.

[83] Galdi, A., Giordano, F., Sposito, A. & Vultaggio, M. (1988). Misure geomagnetiche nel Golfo di Pozzuoli: Metodologia e risultati. *Atti del 7° Convegno GNGTS-CNR*, *3*, 1647-1658.

[84] Galluzzo, D. & La Rocca, M. (2013). Wavefront distortion across Mt. Vesuvius observed by a seismic array. *Annals of Geophysics*, *56* (4), S0448, doi:10.4401/ag-6454.

30] Putignano, M. L., Cinque, A., Lozej, A. & Moccheggiani Carpano, C. (2009). Late-Holocene ground movements in the Phlegrean volcanic district (Southern Italy). New geoarcheological evidence from the islands of Vivara and Procida. *Méditerranée, 112* (2009).

31] Ranieri, G. & Mirabile, L. (1991). Ricerca ed applicazione di metodi geofisici al rilievo sperimentale della struttura medio-profonda dell'area flegrea con uso di sorgenti sismiche watergun. *Annali IUN*, Vol. *63*.

32] Reynolds, R. L., Rosenbaum, J. G., Hudson, M. R. & Fishman, N. S. (1990). Rock magnetism, the distribution of magnetic minerals in the Earth's crust and aeromagnetic anomalies. In: *Geologic Applications of modern aeromagnetic surveys*, (Hanna W.F., Ed.), U.S. Geological Survey Bulletin, 1924, 24-45.

33] Rittmann, A. (1930). Geologie der insel Ischia. *Ergn vur Vulk*, Berlin, Germany.

34] Rosi, M. & Sbrana, A. (1987). Phlegrean Fields. Quaderni De La Ricerca Scientifica, *CNR*, Vol. *114*, No 9, 175 pp.

35] Saccorotti, G., Ventura, G. & Vilardo, G. (2002). Seismic swarms related to diffusive processes: the case of Somma-Vesuvius volcano, Italy. *Geophysics, 67*, 199-203.

36] Santacroce, R. (1987). Somma-Vesuvius. CNR, Quaderni de La Ricerca Scientifica, Roma, Italy.

37] Scandone, P. (1979). Origin of the Tyrrhenian sea and Calabrian Arc. *Bollettino della Società Geologica Italiana, 98*, 27-34.

38] Scandone, R. & Giacomelli, L. (2013). Volcanic precursors in light of eruption mechanisms at Vesuvius. *Annals of Geophysics*, *54*, doi: 10.4401/ag-6461.

39] Scarpa, R., Tronca, F., Bianco, F. & Del Pezzo, E. (2002). High-resolution velocity structure beneath Mount Vesuvius from seismic array data. Geophysical Research Letters, 29, 2040, doi:10.1029/2002GL015 576.

40] Scarpati, C., Cole, P. & Perrotta, A. (1993). The Neapolitain Yellow Tuff – a large volume multiphase eruption from Campi Flegrei, Southern Italy. *Bulletin of Volcanology*, *55*, 343-356.

41] Secomandi, M., Paoletti, V., Aiello, G., Fedi, M., Marsella, E., Ruggieri, S., D'Argenio, B. & Rapolla, A. (2003). Analysis of the magnetic anomaly field of the volcanic district of the Bay of Naples, Italy. *Marine Geophysical Researches*, *24*, 207-221.

42] Siniscalchi, A., Angelino, A., Ruggieri, S., Aiello, G. & Marsella, E. (2002). High resolution magnetic anomaly map of the Bay of Naples (Southern Tyrrhenian sea, Italy). *Bollettino di Geofisica Teorica Applicata*, *42*, 99-104.

143] Thouret, J. C. (1999). – Volcanic geomorphology – an overview. *Earth Science Reviews*, *47*, 95-131.

144] Tibaldi, A. & Vezzoli, L. (2004). A new type of volcano flank failure: the resurgent caldera sector collapse, Ischia, Italy. *Geophysical Research Letters*, *31*, do 10.1029/2004GL020419.

145] Todesco, M., Neri, A., Esposti Ongaro, T., Papale, P., Macedonio, R. & Santacroce (2002). Pyroclastic flow hazard at Vesuvius from numerical modeling. *Bullet Volcanology*, *64*, 155-177.

146] Vail, P. R., Todd, R. G. & Sangree, J. B. (1977). Seismic stratigraphy and changes of sea level. *AAPG Mem.*, *26*, 99-116.

147] Vezzoli, L. (1988) Island of Ischia. CNR, Quaderni De La Ricerca Scientific Italy.

[85] Ghiara, M. R., Lirer, L. & Munno, R. (1979). Mineralogy and geochemistry of the low potassium series of the Campania volcanics (South Italy). *Chemical Geology*, *26*, 29-49.

[86] Gillot, P. Y., Chiesa, S., Pasquarè, G. & Vezzoli, L. (1982). 33.000 y. K/Ar dating of the volcano-tectonic horst of the isle of Ischia, Gulf of Naples. *Nature*, *229*, 242-245.

[87] Ippolito, F., D'Argenio, B., Pescatore, T. S. & Scandone, P. (1973). Unità stratigrafico-strutturali e schema tettonico dell'Appennino meridionale. Istituto di Geologia e Geofisica, Università di Napoli, Pubblicazione n. 15.

[88] Jones, E. J. W. (1999). Marine Geophysics. Wiley, 474 pp., ISBN: 978-0-471-98694-2.

[89] Kennedy, G. C. & Holser, W. T. (1966). Pressure-volume temperature and phase relations od water and carbon dioxide. In: *Handbook of physical constants* (Clark S.P., Ed.), *Geological Society of America Mem.*, *97*, 371-383.

[90] Knott, S. T. & Hersey, J. B. (1956). Interpretation of high resolution echo-sounding techniques and their use in bathymetry, marine geophysics and biology. *Deep Sea Research*, *4*, 36-44.

[91] Latmiral, G., Segre, A., Bernabini, M. & Mirabile, L. (1971). Prospezioni sismiche per riflessione nei Golfi di Napoli e Pozzuoli ed alcuni risultati geologici. *Bollettino della Società Geologica Italiana*, *90*, 163-172.

[92] Lima, A., Plant, J. A., De Vivo, B., Tarvainen, T., Albanese, S. & Cicchella, D. (2008). Interpolation methods for geochemical maps: A comparative study using arsenic data from European steam waters. *Geochemistry, Exploration, Environment, Analysis*, *8* (1), 41-48.

[93] Lomax, A., Zollo, A., Capuano, P. & Virieux, J. (2001). Precise, absolute earthquake location under Somma-Vesuvius volcano using a new three dimensional velocity model. *Geophysical Journal International*, *146*, 313-331.

[94] Marianelli, P., Metrich, N. & Sbrana, A. (1999). Shallow and deep reservoirs involved in magma supply of the 1944 eruption of Vesuvius. *Bulletin of Volcanology*, *61*, 48-63.

[95] Mariani, M. & Prato, R. (1988). I bacini neogenici costieri del margine tirrenico: approccio sismico-stratigrafico. *Memorie della Società Geologica Italiana*, *41*, 519-531.

[96] Masci, F., Palangio, P. & Di Persio, M. (2008). The INGV tectonomagnetic network. *Advances in Geosciences*, *14*, 65-68.

[97] Mastrolorenzo, G., Palladino, D., Vecchio, G. & Taddeucci, J. (2002). The 472 A.D. Pollena eruption at Somma-Vesuvius (Italy) and its environmental impact at the end of Roman Empire. *Journal of Volcanology and Geothermal Research*, *113*, 19-36.

[98] Milia, A. & Giordano, F. (2002). Holocene stratigraphy and depositional architecture of eastern Pozzuoli Bay (eastern Tyrrhenian sea margin, Italy): the influence of tectonics and wavsue-induced currents. *Geomarine Letters*, *22*, 42-50.

[99] Milia, A. & Torrente, M. M. (2007). The influence of paleogeographic setting and crustal subsidence on the architecture of ignimbrites in the Bay of Naples (Italy). *Earth and Planetary Science Letters*, *263*, 192-206.

[100] Mirabile, L. (1969). Prime esperienze di stratigrafia sottomarina eseguite presso l'Istituto Universitario Navale. *Annali IUN*, Vol. *38*.

[101] Mirabile, L., Nicolich, R., Piermattei, R. & Ranieri, G. (1989). Identificazione delle strutture tettonico-vulcaniche dell'area flegrea: sismica multicanale del Golfo di Pozzuoli. *Atti del 7° Convegno GNGTS*, *2*, 829-838.

[102] Mirabile, L., Fevola, F., Galeotti, F., Ranieri, G. & Tangaro, G. (1991). Sismica monocanale ad alta risoluzione con sorgente multi spot di tipo sparker: applicazione ai dati di tecniche di deconvoluzione. *Atti del 10° Convegno GNGTS*, Roma, Italy.

[103] Mirabile, L., De Marinis, E. & Frattini, M. (2000). The Phlegrean Fields beneath the sea: the underwater volcanic district of Naples, Italy. *Bollettino di Geofisica Teorica Applicata*, *41* (2), 159-186.

[104] Mitchum, R. M., Vail, P. R. & Thompson, S. (1977). Seismic stratigraphy and global changes of sea level, part 2: the depositional sequence as a basic unit for stratigraphic analysis. In: *Seismic Stratigraphy – Applications to Hydrocarbon Exploration*, (Payton C.E., Ed.), *AAPG Mem*., *26*, 53-62.

[105] Molin, P., Acocella, V. & Funiciello, R. (2003). Structural, seismic and hydrothermal features at the border of an active, intermittent resurgent block: Ischia island (Italy). *Journal of Volcanology and Geothermal Research*, *121*, 65-81.

[106] Monti, L., D'Argenio, B., Perrotta, A., Scarpati, C., Merola, D., Insinga, D., Lepore, S., Marsella, E., Putignano, M. L., Orrù, P., Aiello, G., Sgrosso, A., Vecchio, E., Conforti, A., De Lauro, M., Di Martino, G., D'Isanto, C., Innangi, S, Passaro, S., Ruggieri, S., Sacchi, M., Tonielli, R., Ferraro, L., Capodanno, M., Molisso, F. & Pelosi, N. (2012). *Foglio geologico 464 Isola di Ischia della Carta 1:25.000 dell'I.G.M.* Regione Campania, Settore Difesa Suolo, ISPRA, Cartografia geologica (scala 1:50.000).

[107] Moore, J. G. & Mark, R. K. (1992). Morphology of the island of the Hawaii. *GSA Today*, *2*, 257-262.

[108] Morhange, C., Marriner, N., Laborel, J., Micol, T. & Oberlin, C. (2006). Rapid sea level movements and non eruptive crustal deformation in the Phlegrean Fields caldera, Italy. *Geology*, *34*, 93-96.

[109] Nabighian, M. N., Grauch, V. J. S., Hansen, R. O., Lafehr, T. R., Peirce, J. W., Phillips, J. D., Ruder, M. E. (2005). The historical development of the magnetic method in exploration. *Geophysics*, *70* (6), 33-61.

[110] Nunziata, C. & Rapolla, A. (1987). A gravity and magnetic study of the volcanic island of Ischia (Naples, Italy). *Journal of Volcanology and Geothermal Research*, *31*, 333-344.

[111] Nunziata, C., Mele, R. & Natale, M. (1999). Shear wave velocities and primary influencing centers of Campi Flegrei-Neapolitan deposits. *Engineering Geology*, *54*, 299-312.

[112] Nunziata, C., Natale, M., Luongo, G. & Panza, G. (2006). Magma reservoir at Mt. Vesuvius: size of the hot, partially molten, crust material detected deeper than 8 km. *Earth and Planetary Science Letters, 242*, 51-57.

[113] Orazi, M., D'Auria, L., Tramelli, A., Buonocunto, C., Capello, M., Caputo, A., De Cesare, W., Giudicepietro, F., Martini, M., Peluso, R. & Scarpato, G. (2013). The seismic monitoring network at Mt. Vesuvius. *Annals of Geophysics*, *56* (4), S0450, doi: 10.4401/ag-6456.

[114] Orsi, G. & Chiesa, S. (1988). The uplift of the Mt. Epomeo block at the island of Ischia (Gulf of Naples): geological and geochemical constraints. *Eos*, *69* (44), 1473.

[115] Orsi, G., Gallo, G. & Zanchi, A. (1991). Simple shearing block-resurgence in caldera depressions. A model from Pantelleria and Ischia. *Journal of Volcanology and Geothermal Research*, *47*, 1-11.

[116] Orsi, G., De Vita, S., Di Vito, M., Isaia, R., Nave, R. & Heiken, (volcanic and related hazard in the Neapolitain area. In: *Earth scie* (Heiken G., Fakundiny R., Sutter J. Eds.), American Geophysica Publication, Washington, pp. 121-170.

[117] Orsi, G., Di Vito, M. A., Selva, J. & Marzocchi, W. (2009). Long eruption style and size at Campi Flegrei caldera (Italy). *Earth and F Letters*, *287*, 265-276.

[118] Ortolani, F. & Aprile, F. (1978). Nuovi dati sulla struttura della Piana est del fiume Volturno. *Bollettino della Società Geologica Italiana*, *97*,

[119] Paoletti, V., Fedi, M., Florio, G., Supper, R. & Rapolla, A. (2004). Th aeromagnetic map of the Phlegrean Fields volcano and surrounding *Geophysics*, *47* (5), 1569-1580.

[120] Paoletti, V., Supper, R., Chiappini, M., Fedi, M., Florio, G. & Rap Aeromagnetic survey of the Somma-Vesuvius volcanic area. *Annals o* (2), 199-213.

[121] Pappalardo, L., Piochi, M. & Mastrolorenzo, G. (2004). The 3550 yr magma plumbing system of Somma-Vesuvius: constraints on its behavi state through a review of Sr-Nd isotope data. *Annals of Geophysics*, *47* (

[122] Parascandola, A. (1924). I crateri dell'isola di Procida. *Bollettino Societ Napoli*, *36*, 57-60.

[123] Perrotta, A., Scarpati, C., Luongo, G. & Morra, V. (2010). St volcanological evolution in the southwestern sector of Campi Flegr island, Italy. In: *Stratigraphy and Geology of Volcanic Areas*, (Gropp L.V. Eds.), GSA Special Papers, ISBN:9780813724645.

[124] Pescatore, T. S., Diplomatico, G., Senatore, M. R., Tramutoli, M. & (1984). Contributi allo studio del Golfo di Pozzuoli: aspetti stratigrafic *Memorie della Società Geologica Italiana*, *27*, 133-149.

[125] Pingue, F., Bottiglieri, M., Godano, C., Obrizzo, F., Tammaro, U., Esposi C. (2013). Spatial and temporal distribution of vertical ground move Vesuvius in the period 1973-2009. *Annals of Geophysics*, *56* doi:10.4401/ag-6457.

[126] Piochi, M., De Vivo, B. & Ayuso, R. A. (2006). The magma feeding syste Vesuvius (Italy) strato-volcano: new inferences from a review of geoche Nd, Pb and O isotope data. In: *Volcanism in the Campania Plain: Ves Flegrei, Ignimbrites*, (De Vivo B., Ed.), *Developments in Volcanology*, 202.

[127] Poli, S., Chiesa, S., Gillot, P. Y., Gregnanin, A. & Guichard, F. (198 versus time in the volcanic complex of Ischia (Gulf of Naples, Italy). *Co Mineralogy and Petrology*, *95* (3), 322-335.

[128] Poli, S., Chiesa, S., Gillot, P. Y., Guichard, F. & Vezzoli, L. (1989). Time the geochemical approach and hazard estimation of a volcanic area: the case (Italy). *Journal of Volcanology and Geothermal Research*, *36*, 327-33

[129] Principe, C., Rosi, M., Santacroce, R. & Sbrana, A. (1987). Explanatory geological map of Vesuvius. In: *Somma-Vesuvius* (Santacroce R., Ed.), *CN De La Ricerca Scientifica*, 11-51.

[148] Wielandt, E. (1975). Generation of seismic waves by underwater explosions. *Geophysical Journal of Astronomical Society*, *40*, 421-439.

[149] Zollo, A., Gasparini, P., Virieux, J., Le Meur, H., et al. (1996) Seismic evidence for a low velocity zone in the upper crust underneath Mt. Vesuvius. *Science*, *274*, 592-594.

[150] Zollo, A., D'Auria, L., De Matteis, R., Herrero, A., Virieux, J. & Gasparini, P. (2002). Bayesian estimation of 2D P-velocity models from active seismic arrival time data imaging of the shallow structure of Mt. Vesuvius (Southern Italy). *Geophysical Journal International*, *151*, 566-582.

[151] Zuppetta, A., Sava, A. & Zuppetta, C. (1993). Evoluzione vulcano-tettonica dell'isola d'Ischia: un modello per gli ultimi 35 ka di attività. *Bollettino della Società Geologica Italiana*, *112* (1), 353-369.

In: Geophysics
Editor: Gemma Aiello

ISBN: 978-1-63484-831-2

Chapter 4

OCEANIC OSCILLATION PHENOMENA: RELATIONSHIPS WITH SYNCHRONIZATION AND STOCHASTIC RESONANCE

Shinya Shimokawa[1] and Tomonori Matsuura[2]
[1]National Research Institute for Earth Science and Disaster Prevention, Tsukuba, Ibaraki, Japan
[2]University of Toyama, Toyama, Toyama, Japan

ABSTRACT

Various oscillation phenomena, such as the transitions between stable and unstable states of western boundary currents, i.e., the Kuroshio, are present in the oceanic circulation of the North Pacific Ocean. These phenomena are considered to be related to non-linear rhythmic phenomena, such as the synchronization and the stochastic resonance, which are often observed in non-linear ocean systems. Synchronization is an adjustment (e.g., frequency and/or phase locking) of the rhythms of two or more self-sustained oscillating systems that have different periods because of their non-linear interaction. Stochastic resonance is a phenomenon in which parts of potential signals can exceed the threshold when adequate noise is added to non-linear systems; these can then be detected as actual signals. However, oceanic applications of these non-linear rhythmic phenomena have not been investigated in detail. Thus, we investigated the responses of oceanic double gyres to external wind forcing with and without noise using a 1.5 layer quasi-geostrophic model and considered the possibility of the occurrence of these non-linear rhythmic phenomena in general oceanic circulation.

Keywords: oceanic circulation, non-linear system, synchronization, stochastic resonance

INTRODUCTION

The Dutch physicist Christiaan Huygens (1629–1695), who is famous for his work on optics, was also the inventor of the pendulum clock. He found that two pendulum clocks mounted on a wall gradually synchronized even when their pendulums had initially been swinging separately. In the time of Huygens, the mechanism causing this phenomenon was unknown, but now, based on modern non-linear science, it can be understood as the adjustment of the rhythms of two or more self-sustained periodic oscillators that have different time periods because of their non-linear interaction [1]. The adjustment can be described in terms of phase locking and frequency entrainment. The non-linear interaction can be achieved in several ways. For example, in the case of two pendulum clocks mounted on a wall, the weak transfer of their oscillation energies through the wall can cause the clocks to adjust and compromise their pace mutually, resulting in synchronization. In the case with a fixed-period oscillator, only another forced oscillator can compromise with the fixed one [2]. The circadian rhythm, in which the active periods of living things on the earth synchronize with the self-rotation period of the earth, is one example of this phenomenon.

Benzi et al. (1981) [3] suggested the concept of stochastic resonance, in which a noise (or disturbance) can amplify weak signals, to explain the observed glacial variability during the Pleistocene, having a period of 100,000 years, which is considered to be related to small variations in the solar orbit. In non-linear systems, no output can be obtained unless the input to the system exceeds a threshold. When a suitable noise that is not excessively weak or strong is added to the system, parts of potential signals are then able to exceed the threshold and can be detected as real signals. Stochastic resonance is a physical phenomenon in which a type of noise amplifies and elicits weak signals that are below the detection threshold. However, when too much noise is added to the system, the potential signals become concealed by the noise. Moreover, if extremely weak noise is added, the potential signals cannot exceed the threshold. Therefore, stochastic resonance shows the interesting characteristic of the existence of an optimum noise strength.

Both synchronization and stochastic resonance are general characteristics of highly non-linear open systems and have been observed in living systems and electrical circuits [4-7]. They are considered to be related to formation of ordered structure, such as in the rhythm adjustment observed in non-linear systems. Thus, its various geophysical applications can be expected. For example, the Moon always shows the same surface to the Earth because the moon is in synchronous rotation with the Earth, which can affect geophysical phenomena such as tide and earthquake [8]. Moreover, stochastic resonance, in the first place, was found in relation to paleoclimate change, and can be related to various present climatological phenomena [9].

The ocean system can be regarded as a highly non-linear open system driven by external forcing (e.g., the wind stress on the sea surface). In the oceanic system, non-linearity emerges mathematically from advection terms in the momentum equation and physically from surplus relative vorticity in the ocean. The ocean system is an open system in the sense that the internal order is maintained by the input of wind stress energy from the external atmosphere. Specifically, in the mid-latitude North Pacific Ocean, a westerly wind blows eastward around 45° N and a trade wind blows westward around 15° N. The water level rises around 30° N because of the southward Ekman transport by the westerly wind and northward Ekman transport by the trade wind. An oceanic current, called a (clockwise) sub-tropical gyre, is generated as a result of the geostrophic balance, causing a sea level rise in the ocean. At the same way, an oceanic current, known as a (counterclockwise) sub-polar gyre, is generated by the balance between the westerly and circumpolar winds. The combination of both gyres is referred to as the oceanic double-gyre and is one of the basic features of large-scale surface oceanic circulation (Figure 1). The variability in the oceanic general circulation results in a non-linear interaction between the atmospheric wind stress oscillator and the oceanic circulation oscillator. In this case, the interaction can be realized between a fixed-period oscillator and a oscillator forced by it, such as the above mentioned circadian rhythm. In fact, previous studies [10-13] have noted the possibility that both the synchronization and the stochastic resonance should play an important role in causing the variability of strong oceanic currents and of the concurrent eddy shedding.

Under a steady wind stress, equivalent barotropic wind ocean models show that transition to chaos through oscillatory instabilities of the mean flow [14] and occurrence of the low-frequency variability associated with a homoclinic bifurcation [15]. Sakamoto (2006) [16] showed that seasonal wind forcing generates long-term variability through a path toward chaos with a quasi-devil's staircase. However, these studies didn't focus directly on synchronization though they seem to be related to it. Shimokawa and Matsuura (2010) [10] investigated the response of oceanic double-gyre to seasonal external forcing and found that synchronization emerges with increasing amplitude of the forcing due to non-linear interactions between the intrinsic oscillation of eddy shedding and seasonal oscillation of external forcing.

Sura et al. (2001) [17] investigated the response of oceanic double-gyre to constant external forcing with white noise. Pierini (2010, 2011) [18, 19] investigated the response of oceanic double-gyre to constant and time-dependent external forcing with red noise (noise with time correlation). However, there had been no sufficient investigations on basic points such as the amplitude of noise added to the system and flow fields when stochastic resonance occurs, and behavior when large noise is added to the system. Shimokawa et al. (2013, 2014) and Shimokawa and Mtasuura (2015) [10-13] focused on and discussed these matters.

One aim of this study is to understand the relevance of synchronization and stochastic resonance in the generation of oceanic turbulence. In this chapter, we introduce new findings on the topic, mainly by the authors.

NUMERICAL MODEL AND METHODS

The model used in this study was a 1.5 layer, reduced-gravity, quasi-geostrophic numerical model with Rayleigh-type interfacial friction and a nonslip boundary condition ([20]; Figure 1). In this model, the vertical structure is assumed to consist of two immiscible (i.e., incapable of mixing) homogeneous layers of slightly different densities. The lower layer is assumed to be infinitely deep and at rest. The boundary between the lower and upper layers corresponds to the oceanic pycnocline.

The governing equation of the model is a standard second-order finite-difference approximation to the Q-G vorticity equation for the upper layer flow, i.e.,

$$(\nabla^2-\gamma^2)\partial h/\partial t + \beta\partial h/\partial x$$
$$= -J(h, \nabla^2 h) - r\nabla^2 h + A_h\nabla^4 h - A_b\nabla^6 h + f_0/\rho_0 g' H_0 \nabla\times\tau \qquad \text{(Eq. 1)}$$

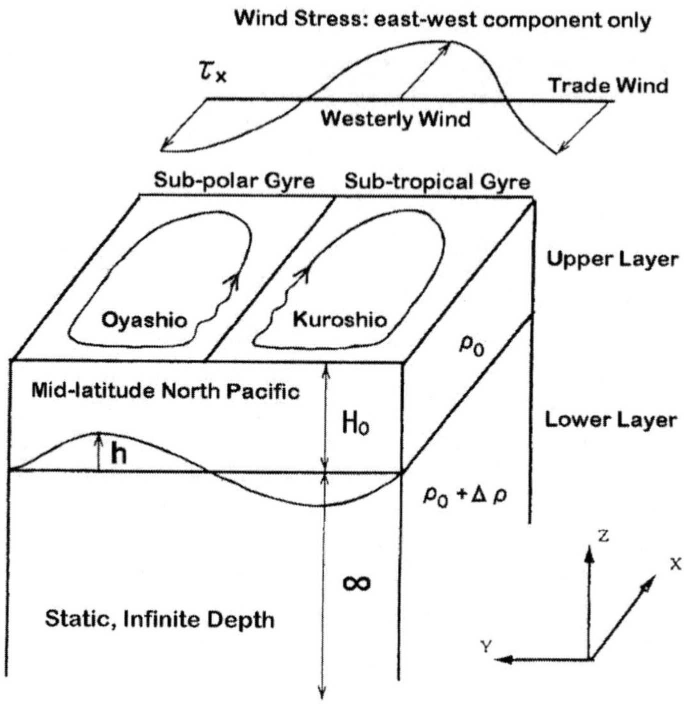

Figure 1. Schematic of numerical model. The vertical structure is assumed to consist of two immiscible (i.e., incapable of mixing) homogeneous layers of slightly different densities (i.e., $\Delta\rho$). H_0 is the reference thickness of the upper layer. The lower layer is assumed to be infinitely deep and at rest. The boundary between the lower and upper layers corresponds to the oceanic pycnocline, and the interface anomaly is represented by h. See also text in the 4th paragraph in the introduction.

where h is the interface anomaly (positive upwards), r is the coefficient of internal friction, A_h is the coefficient of eddy viscosity, A_b is the coefficient of high-order (bi-harmonic) viscosity, τ is the wind stress on the sea surface, ρ_0 is the mean density of seawater, and H_0 is the reference thickness of the upper layer, t is the integration time, and x is the location in the E-W direction. ∇ and J are the Nabla and Jacobian operators, respectively. Other definitions include

$$\gamma^2 = f_0^2/g'H_0 \qquad \text{(Eq. 2)}$$

$$g' = g\Delta\rho/\rho_0 \qquad \text{(Eq. 3)}$$

and

$$\beta = \partial f/\partial y \qquad \text{(Eq. 4)}$$

Here, f_0 is the value of the Coriolis parameter at the middle of the domain, g is the gravitational acceleration, and $\Delta\rho$ is the difference between the layer densities, y is the location in the N-S direction, and f is the Coriolis parameter (assumed to be a function of y). As fixed parameters, $A_b = 0.0$ m^4 s^{-1}, r = 1.0×10^{-7} s^{-1}, $f_0 = 7.3 \times 10^{-5}$ s^{-1}, $\beta = 2.0 \times 10^{-11}$ m^{-1} s^{-1}, $g' = 2.0 \times 10^{-2}$ m s^{-2}, and $H_0 = 1000$ m are used.

The model domain was a rectangular area of 3600 km × 2800 km (representing the mid-latitude North Pacific), and the horizontal grid spacing was 2.0×10^4 m (20 km). The integration period was 250 years, and the time interval of integration was 7.2×10^3 s (2 h). The velocity of the western boundary currents, such as the Kuroshio current, was assumed to be approximately 0.1–1.0 m s^{-1}, and the diameter of mesoscale eddies was taken to be approximately 10–100 km. Considering these values, the effect of sub-grid scale eddies was represented by the horizontal eddy viscosity, and the coefficients, A_h, were selected from the range 10^2–10^3 m^2 s^{-1}. Further details of the numerical model are provided in [10].

The external wind forcing, $\tau(y, t)$, has the following form:

$$\tau(y, t) = [1 + \alpha \cos(\omega t) + \varepsilon\, n(t)/\sigma_n] \times \tau_0 \cos(2\pi y/L), \qquad \text{(Eq. 5)}$$

where α is the amplitude of seasonal variation (see below in this section), ω is the frequency of seasonal variation (= 1/1 $year^{-1}$), ε is the amplitude of red noise $n(t)$ (see below in this section), σ_n is the root mean square of $n(t)$, τ_0 is the amplitude of wind stress (= 0.1 N m^{-2}), and L is the length of the region in the N-S direction (= 2800 km). The basic external forcing field, $\tau_0 \cos(2\pi y/L)$, has only an E-W component, is constant in time and varies in the N-S direction, representing a simplified spatial distribution of wind stress in the mid-latitude North Pacific. Seasonal variation and red noise were added to the basic field after year 50 in the integration.

The red noise, $n(t)$, is generated by the Ornstein–Uhlenbeck equation [21, 18] using the following expressions:

$$dn/dt = -an + b\eta \qquad \text{(Eq. 6)}$$

and

$$n_{k+1} = cn_k + d\eta_k, \qquad \text{(Eq. 7)}$$

where η is white Gaussian noise with zero mean and unit variance, and a and b are positive constants. In addition, c = (1 − a Δt) and d = b Δt, where Δt is the integration time interval in Eq. (7) (which is the first-order autoregressive model used to solve Eq. (6) numerically), and k is the time step. In this study, we have used constant values of b = 1 and Δt = 1 day. Therefore, the only parameter related to red noise is a (or c). The parameter a is related to the decorrelation time T_S by the equation

$$T_S = 1/a = \Delta t/(1 - c) \qquad \text{(Eq. 8)}$$

In this study, 1 year was used as T_S, (i.e., a = 1/1 $year^{-1}$), and σ_n was assumed to be a constant value because x represented a stationary process. In this model, a = 0 and a = 1 respectively correspond to Gaussian white noise and to a Wiener process. Further details of the red noise calculation are provided in [12].

Figure 2 shows the Reynolds number (Re) dependence of changes in the total energy (the sum of kinetic and available potential energies) for constant external forcing experiments (i.e., $\alpha = \varepsilon = 0.0$). In these cases, Re changes only with changes in the horizontal eddy viscosity coefficient, A_h (Re increases with decreasing A_h).

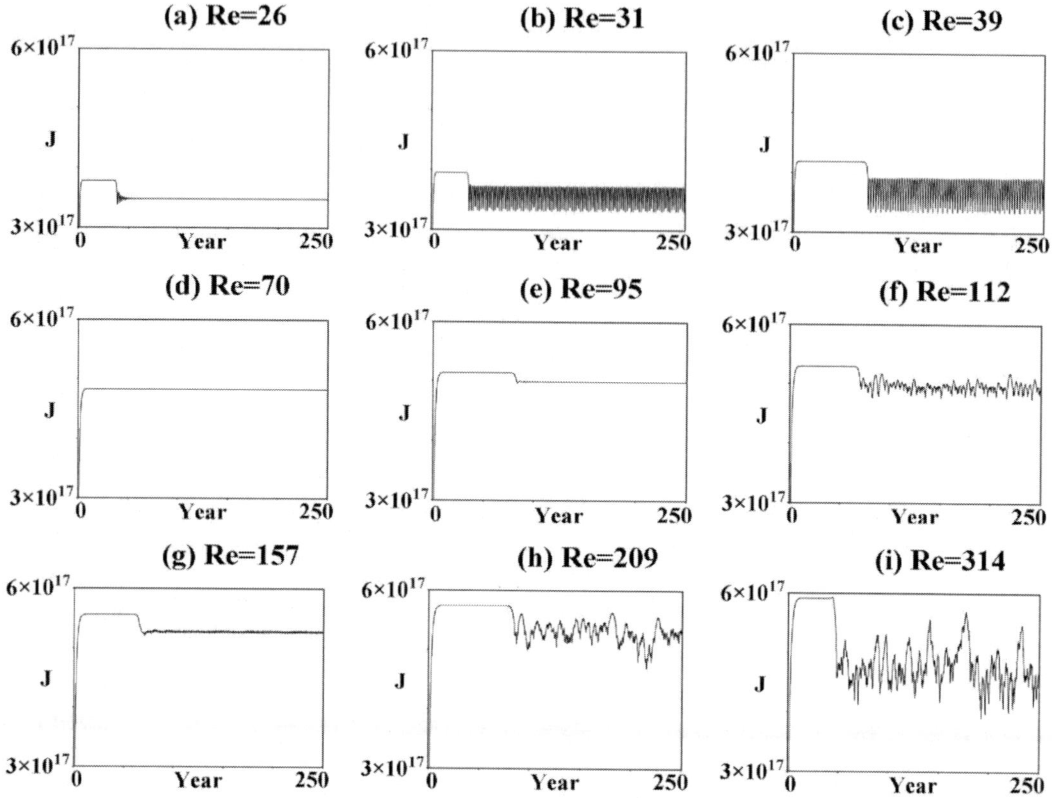

Figure 2. Time series of total energy in experiments under time-independent forcing for Re = (a) 26, (b) 31, (c) 39, (d) 70, (e) 95, (f) 112, (g) 157, (h) 209, and (i) 314 (after [11]).

For Re = 26 (Figure 2a), the energy is stable; for Re = 31 and 39 (Figures 2b and 2c), it begins to oscillate regularly; and for Re = 70 and 95 (Figures 2d and 2e), the energy becomes stable again. For Re = 112 (Figure 2f), the energy begins to oscillate irregularly; for Re = 157 (Figure 2g), it returns to stability; and for Re = 209 and 314, the energy begins to oscillate irregularly and with a large amplitude. The changes can be classified into four groups as it follows: 1) Re = 26, 31, 39; 2) Re = 70, 95, 112; 3) Re = 157, 209; and 4) Re = 314.

Each group has a similar average energy level (after the first rapid changes for each Re, for example, after around year 38 for Re = 26; see Figure 2a). In other words, the energy level changes discontinuously (i.e., quantization with respect to energy level occurs) and represents a different quality state. As a result, the state of Re = 39 represents the state in which the modelled ocean regularly oscillates with a characteristic period. The state with Re = 157 represents the situation in which potential signals exist in the model ocean but are not apparent.

Some details of the conducted experiments are provided below.

[Seasonally changing experiments] These experiments were conducted under seasonally varying external forcing with a 1 year period for Re = 39. In these experiments, the only variable parameter was α, the amplitude of the seasonal variation (i.e., $\varepsilon = 0.0$), which can vary from 0.0 to 1.0: $\alpha = 0.0$ corresponds to no seasonal variation, and $\alpha = 1.0$ indicates that the amplitude of the seasonal variation is equal to that of the average external force.

[Noise-added experiments] These experiments were conducted under red-noise-added external forcing for Re = 150. In these experiments, the only variable parameter was ε, the amplitude of the seasonal variation (i.e., α = 0.0), which can vary from 0.0 to 0.1: ε = 0.0 corresponds to no added red noise, and ε = 0.25 indicates that the amplitude of the red noise is equal to that of the average of external force because the range of $x(t)/\sigma_x$ in Eq. (1) is roughly equal to ±4.0 [18].

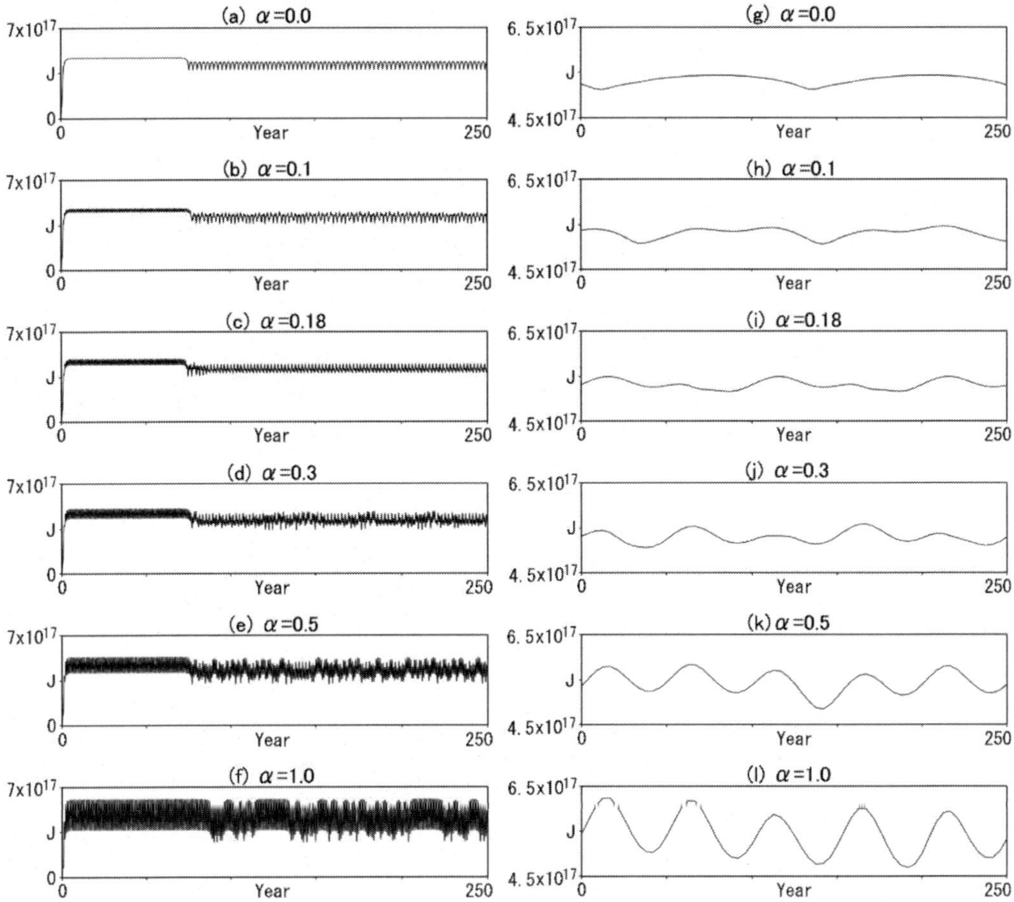

Figure 3. Time series of total energy in seasonally changing experiments over 250 years for α = (a) 0.0, (b) 0.1, (c) 0.18, (d) 0.3, (e) 0.5, and (f) 1.0 and over 5 years (from year 100 to year 105) for α = (g) 0.0, (h) 0.1, (i) 0.18, (j) 0.3, (k) 0.5, and (l) 1.0 (after [10]).

Here, the physical meaning of the addition of noise to wind stress and the corresponding decorrelation time are discussed. The addition of noise to wind stress corresponds to the random components of the wind stress in the real atmosphere. Our study focused on whether new variation could occur as a result of the effects of wind stress noise on the mechanism of stochastic resonance in the ocean. In addition, we assumed that the random component is related to past atmospheric states to some extent and could be represented by red noise rather than by white noise, which has no relationships with the past. However, the adequate decorrelation time is not clear. We assumed that the decorrelation time was 1 year because a 1 year periodic component is dominant in the wind stress in the real atmosphere; however, this point requires further investigation [18, 19, 12]. In addition, we assumed that the spatial distribution of noise is uniform; this point would also benefit from further study [17].

RESULTS

1.1. Seasonally Changing Experiments and Synchronization

Figure 3 shows the Re dependence of changes in the total energy in the seasonally varying experiments (i.e., $\varepsilon = 0.0$). When $\alpha = 0.0$ (i.e., no seasonal variation), only the characteristic oscillation of the ocean can be observed (Figures 3a and 3g). When $\alpha = 1.0$ (i.e., the amplitude of seasonal variation is large), only the seasonal oscillation (with a period of 1 year) due to the external forcing can be observed (Figures 3f and 3l). When $\alpha = 0.18$ (i.e., the amplitude of seasonal variation is intermediate), the oscillation shows both aspects (Figures 3c and 3i). In this case, the synchronization occurs, as it will be described later.

With increasing α, intermittency (irregular repetition of static states and turbulent states) appears ($\alpha > 0.4$; Figures 3e and 3f). The periods of intermittency range from a few years to a few decades. Although the intermittency in cases with large α is an interesting phenomenon that is considered to be related to irregular oceanic variations (e.g., transitions between a stable modon-like pattern and unstable cyclonic/anti-cyclonic eddy shedding patterns [10]); in this report, we focus on the synchronization for the case with small α.

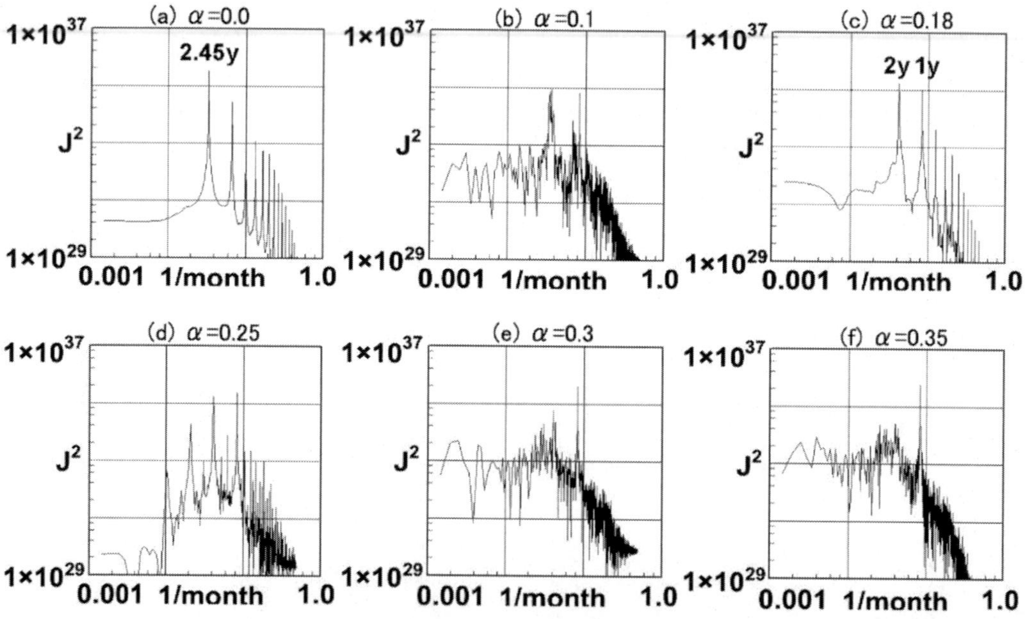

Figure 4. Power spectra of total energy in seasonally changing experiments for α = (a) 0.0, (b) 0.1, (c) 0.18, (d) 0.25, (e) 0.3, and (f) 0.35 (after [10]).

Figure 4 shows the total energy power spectra in the seasonally changing experiments for $\alpha < 0.4$. When $\alpha = 0.0$ (Figure 4a), only the characteristic period of the model ocean (i.e., 2.45 years) can be observed. With increasing α (Figure 4b), the spectrum becomes turbulent. However, when $\alpha = 0.18$ (Figure 4c), a regular pattern suddenly appears in the spectrum. In this case, the characteristic period of the modelled ocean (2.45 years) is synchronized to twice the period of the external forcing (2 years). The period of the whole system is considered to be adjusted by the interactions between the ocean system and the external forcing. This behavior is considered to be an example of synchronization with a fixed-period oscillator (see the introduction). In addition, in this case, random peaks around the peaks of 1 and 2 years are synchronized into these two peaks. When $\alpha = 0.25$ (Figure 4d), a periodic doubling occurs, and with increasing α (Figures 4e and 4f), the spectra become turbulent again.

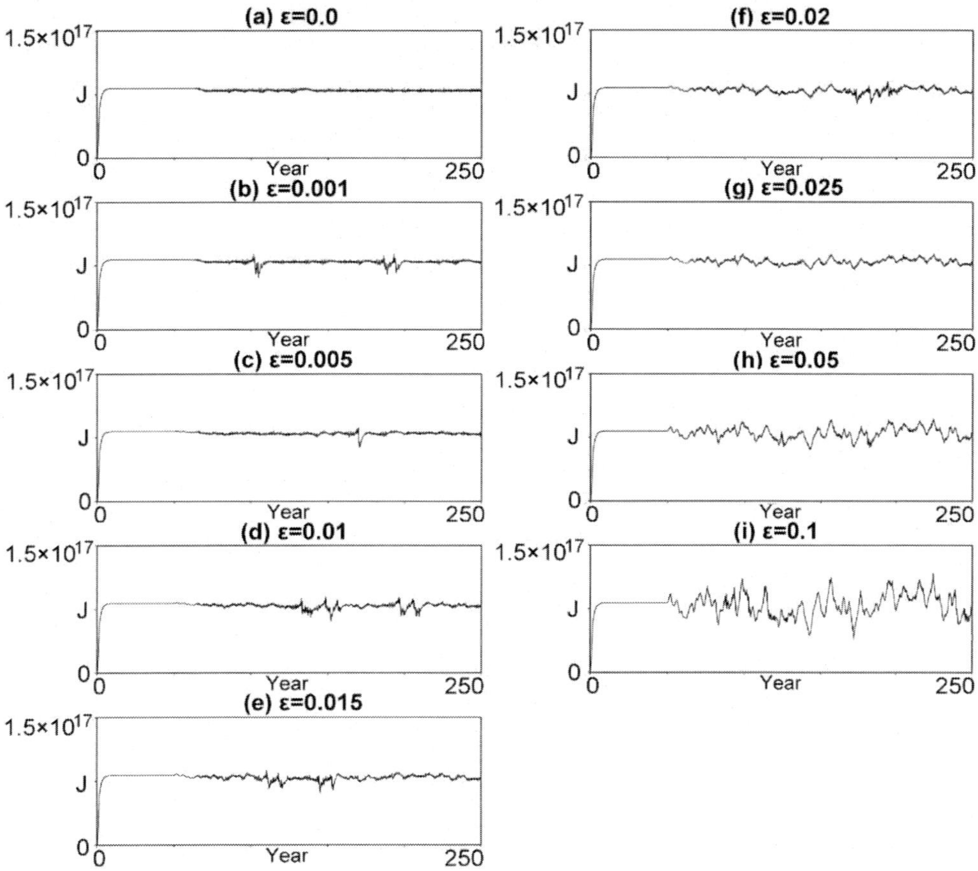

Figure 5. Time series of total energy in noise-added experiments for ε = (a) 0.0, (b) 0.001, (c) 0.005, (d) 0.01, (e) 0.015, (f) 0.02, (g) 0.025, (h) 0.05, and (i) 0.1 (after [12]).

1.2. Noise-Added Experiments and Stochastic Resonance

Figure 5 shows the ε dependence of the kinetic energy variations in the noise-added experiments (i.e., $\alpha = 0.0$). For the case of $\varepsilon = 0.0$ (no noise), a large variation is not evident.

When 0.001 $\leq \varepsilon \leq$ 0.015, certain intermittent large variations are visible. These results indicate that potential signals in the system are amplified by the noise added to the system and appear at the front as actual signals (i.e., stochastic resonances occur). For $\varepsilon \geq$ 0.02, long-term (i.e., slow) variations are observed. The long-term variations show a similar form for all cases with $\varepsilon \geq$ 0.02, although the amplitude of the variations is dependent on ε. Moreover, short-term variations within the long-term variations do not show intermittency, such as when 0.001 $\leq \varepsilon \leq$ 0.015. These results indicate that the system variations are controlled by strong noise and that potential signals in the system are buried in the noise.

When the numbers of the intermittent large variations occurred in Figure 5 are regarded as the numbers of the stochastic resonance occurred, the peak positions of the numbers seem to occur around ε = 0.001 and ε = 0.01. In this discussion, we focus on the neighborhood of ε = 0.001 because that of ε = 0.01 is considered to be affected by strong noise to some extent.

Figure 6 shows the time series of the total energy for the noise-added experiments conducted with ε between 0.0002 and 0.001 in increments of 0.0001. The results indicate that the peak position of the numbers is around ε = 0.0006, and there is an overall trend for the values to decrease with distance for ε higher than 0.0006. These results indicate the existence of an optimum noise strength in stochastic resonance in the parameter range corresponding to a noise that is not excessively strong. However, the numbers are less than 10 for 250 years and can differ significantly over longer integration times. In fact, for ε = 0.003, there are intermittent large variations, and the numbers (and hence the peak position) can vary over longer integration periods. Therefore, to explore this behavior fully, experiments using longer integration times will be required.

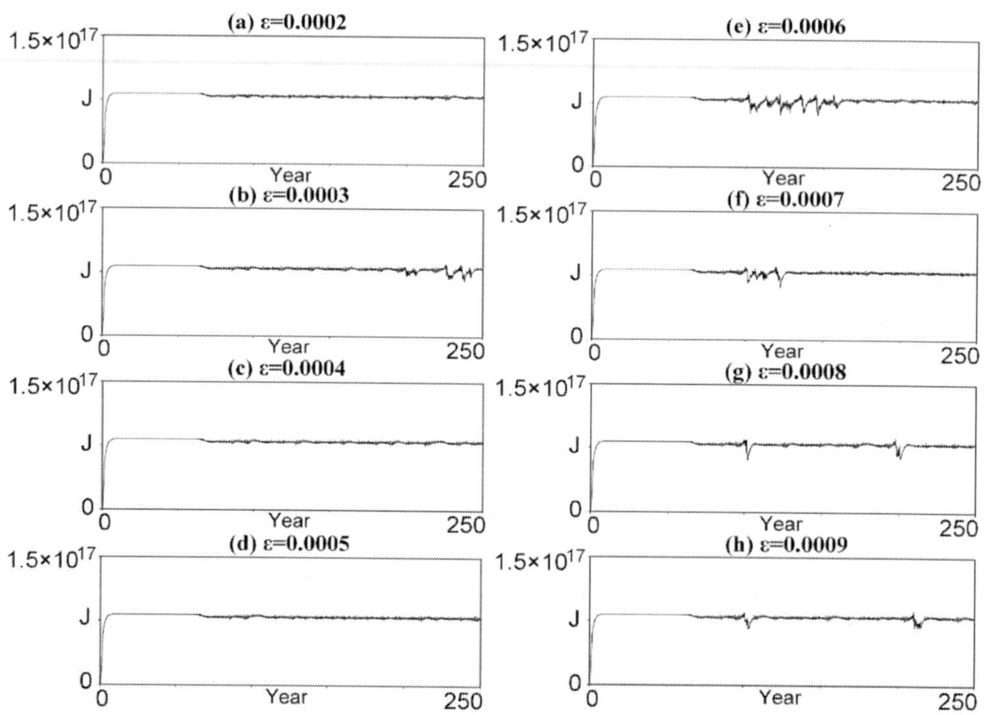

Figure 6. Time series of total energy in noise-added experiments for ε = (a) 0.0002, (b) 0.0003, (c) 0.0004, (d) 0.0005, (e) 0.0006, (f) 0.0007, (g) 0.0008, and (h) 0.0009 (after [13]).

SUMMARY AND DISCUSSION

The results obtained in this study can be summarized as follows.

- The response of the oceanic double-gyre to seasonally changing external forcing was investigated by changing only the seasonal variation amplitude α in the numerical model. We focused on the state Re = 39, in which the modelled ocean regularly oscillated with a characteristic period.
- With seasonally varying external forcing, the characteristic period of the modelled ocean (2.45 years) was synchronized to twice the period of the external forcing (2 years). In this case, the period of the whole system is believed to adjust due to the interactions between the ocean system and the external forcing.
- The response of the oceanic double-gyre to red-noise-added external forcing was investigated by varying only the red noise amplitude ε in the numerical model. We focused on the condition Re = 157, in which potential signals existed in the model ocean but did not appear in the results.
- With the noise-added external forcing, an adequate noise amplified the potential signals in the system (i.e., stochastic resonances occur), but strong noise buried the potential signals within the noise.
- An optimum noise strength in stochastic resonance is considered to exist at least in the parameter range corresponding to a noise that is not excessively strong.

However, it is not clear whether the last result is valid across the whole parameter range. To clarify this point, more detailed experiments are necessary. This characteristic is interesting because it is formally identical to the "intermediate-disturbance hypothesis" [22, 23], that has been demonstrated in ecosystem studies, in which species diversity in a local area is maximized when the environmental disturbances are neither excessively weak nor excessively strong.

Our studies [10-13] have noted also the possibility that both the synchronization and the stochastic resonance significantly influence the variability of strong oceanic currents and of the concurrent eddy shedding. These phenomena may affect the periods of oceanic oscillation phenomena, such as the transitions between stable and unstable states of western boundary currents, e.g., the Kuroshio. The roles of these phenomena in the real ocean should be investigated in detail.

The coupling of synchronization and stochastic resonance has also been discovered and is known as stochastic synchronization [24, 25]. However, only one study [26] related to stochastic synchronization in oceanic circulation has been conducted so far; thus, additional investigations on this topic are required.

ACKOWLEDGMENTS

We thank the American Meteorological Society, National Congress of Theoretical and Applied Mechanics Japan, and Japanese Society of Multi Phase Flow for granting permission to reuse the figures. This study was supported by the Japan Society for the Promotion of

Science through grants No. 25400465 and a research project on coastal disaster by the National Research Institute for Earth Science and Disaster Prevention.

REFERENCES

[1] Kuramoto, Y., 1984, *Chemical Oscillations, Waves, and Turbulence,* Springer, Berlin.

[2] Pikovsky, A., M. Rosenblum and J. Kurths, 2001, *Synchronization: A Universal Concept in Nonlinear Sciences,* Cambridge Univ. Press., Cambridge.

[3] Benzi, R., A. Sutera and A. Vulpiani, 1981, The Mechanism of Stochastic Resonance, *J. Phys.,* 14A, L453-L457.

[4] Moss, F. and K. Wisenfeld, 1995, *The Benefits of Background Noise, Scientific American,* August 1995, 66-69.

[5] Benzi, R., 2010, Stochastic Resonance: from Climate to Biology, *Nonlin. Processes Geophys.,* 17, 431-441.

[6] Kuramoto, Y., 2014, *Nonlinear Science - Synchronizing World* (in Japanese), Syueisha, Tokyo.

[7] Kori, H., 2014, *Perspective on Synchronization (in Japanese),* Butsuri, 69, 602-603.

[8] Tanaka, S., M. Ohtake, and H. Sato, 2002, Evidence for Tidal Triggering of Earthquakes as Revealed from Statistical Analysis of Global Data, *J. Geophys. Res.,* 107 (B10), 2211, doi:10.1029/2001JB001577.

[9] Nicolis, G and I. Prigogin, 1989, *Exploring Complexity: An Introduction,* W. H. Freeman & Co., Hampshire.

[10] Shimokawa, S. and T. Matsuura, 2010, Chaotic Behaviors in the Response of Q-G Oceanic Double-Gyre to Seasonal External Forcing, *J. Phys. Oceanogr.,* 40, 1458-1472.

[11] Shimokawa, S., T. Matsuura and H. Hashimoto, 2013, Instability and Long-Term Variability of Strong Current Extension in an Oceanic Double-Gyre, *Theor. Appl. Mech. Jpn.,* 61, 23-30.

[12] Shimokawa, S., K. Shiratori, and T. Matsuura, 2014, Responses of Oceanic Double-Gyre to External Wind Forcing with Noise, *Theor. Appl. Mech. Jpn.,* 62, 49-57.

[13] S. Shimokawa and T. Matsuura, 2015, Oscillation Phenomena in the Oceanic General Circulation and Rhythmic Phenomena in Nonlinear Systems, *Japanese J. Multiphase flow,* 28, 583-590 (in Japanese).

[14] Dijkstra, H. A., 2000, *Nonlinear Physical Oceanography: A Dynamical Systems Approach to the Large Scale Ocean Circulation and El Nino,* Springer, Dordrecht.

[15] Nagida, B. T. and B. Luce, 2001, Global Bifurcation of Shilnikov Type in a Double-Gyre Model, *J. Phys. Oceanogr.,* 31, 2669-2690.

[16] Sakamoto, T. and I. Umetsu (2006), Seasonal Energy Cycle of Wind-driven Ocean Circulation with Particular Emphasis on the Role of Bottom Topography, *Deep Sea Res.,* I53, 154-168.

[17] Sura, P., K. Fraedrich and F. Lunkeit, 2001, Regime Transitions in a Stochastically Forced Double-Gyre Model, *J. Phys. Oceanogr.,* 31, 411–426.

[18] Pierini, S., 2010, Coherence Resonance in a Double-Gyre Model of the Kuroshio Extension, *J. Phys. Oceanogr.,* 40, 238–248.

[19] Pierini, S., 2011, Low-Frequency Variability, Coherence Resonance, and Phase Selection in a Low-Order Model of the Wind-Driven Ocean Circulation, *J. Phys. Oceanogr.,* 41, 1585-1604.

[20] McCalpin, J. D., 1995, The Statistics and Sensitivity of a Double-Gyre Model: The Reduced-Gravity, Quasigeostrophic Case, *J. Phys. Oceanogr.,* 25, 806–824.

[21] Ornstein, L. S., and Uhlenbeck, G. E., 1930, On the Theory of the Brownian Motion, *Phys. Rev.,* 36, 823-841.

[22] Connell, J. H., 1978, Diversity in Tropical Rain Forests and Coral Reefs, *Science,* 199, 1302-1310.

[23] Shimokawa, S., T. Murakami, A. Ukai, H. Kohno, A. Mizutani, and K. Nakase, 2014, Relationship Between Coral Distributions and Physical Variables in Amitori Bay, Iriomote Island, Japan, *Journal of Geophysical Research: Oceans,* 119, 8336-8356, doi: 10.1002/2014JC010307.

[24] Fukuda, H., H. Nagano, and S. Kai, 2003, Stochastic Synchronization in Two-Dimensional Coupled Lattice Oscillators in the Belousov–Zhabotinsky Reaction, *J. Phys. Soc. Jpn.,* 72, 487-490.

[25] Kai, S., 2005, Rhythm and Stochastic Resonance in Living System (in Japanese), *The World of Rhythmic Phenomena* (Y. Kuramoto ed., University of Tokyo Press, Tokyo), 33-95.

[26] Shimokawa, S. and T. Matsuura, 2015, Stochastic synchronization in an oceanic double gyre, *Theor. Appl. Mech. Jpn.,* 63, in press.

In: Geophysics
Editor: Gemma Aiello

ISBN: 978-1-63484-831-2

Chapter 5

ASSESSMENT OF OCEAN VARIABILITY IN THE SICILY CHANNEL FROM A NUMERICAL THREE-DIMENSIONAL MODEL USING EOFS DECOMPOSITION

R. Sorgente[1,*], A. Olita[1], A. F. Drago[2], A. Ribotti[1], L. Fazioli[1] and C. Tedesco[1]

[1]Institute for Coastal Marine Environment of the National Research Council in Oristano, Loc. Sa Mardini, Oristano, Italy
[2]Physical Oceanography Unit, IOI-Malta Operational Oceanography, University of Malta, B'Kara, Malta

ABSTRACT

The circulation in the Sicily Channel and the surrounding areas has been simulated from January 2001 to December 2004 using a numerical three-dimensional, free surface, nested limited area ocean model.

Basic monthly-based statistics, as well as Empirical Orthogonal Functions (EOF) analysis, have been used to study the seasonal and interannual variability of the circulation and hydrology of the Sicily Channel. Significant interannual variability of the surface properties, superimposed over an underlying and stronger seasonal cycle, has been identified in the space-time domain through the EOFs and has been correlated with the changes in the simulated circulation and hydrology. The main interannual event is related to changes in the surface circulation that occurred during summer 2003. The skill of the numerical model is assessed to measure its ability in reproducing the dynamical characteristics of this part of the Mediterranean, taking into account the general lack of data, especially in the southern part of the domain. The model performances also suggest

* Corresponding author: Roberto Sorgente, Institute for Coastal Marine Environment of the National Research Council in Oristano, Loc. Sa Mardini, 09170 Oristano – Italy. Tel.: +39.0783.22027, Fax +39.0783.22002, E-mail: roberto.sorgente@cnr.it.

directions for further research efforts and for model improvements, such as an improved reproduction of the mixed layer.

Keywords: numerical model, sicily channel, EOF, variability, mediterranean sea

INTRODUCTION

The central Mediterranean region [1] (Figure 1) is characterized by a number of significant dynamical processes covering the full spectrum of temporal and spatial scales [2, 3, 4]. In addition to the general oceanographic circulation with its mesoscale variability, there are the wind-driven currents on the shelf from local and remote storms and upwelling off Sicily [5, 6], and then tidal, inertial, gravity, surface and continental shelf waves [7]. The time variability of the water mass properties, circulation and water transport has been discussed by several authors using in-situ ocean observations [8, 9, 10, 11, 1, 12, 13, 14]. These observations are often sporadic with a coarse horizontal resolution limited to the Italian seas and generally lacking over the North African continental shelf. Moreover, since the Rossby radius over the shelf area is less than 10 km, the mesoscale phenomena with periods from 3 to 10 days [15] cannot often be detected and followed with enough a detail in both time and space domains due to this lack in observations. These constraints do not permit the identification of the main spatial and temporal scales of the sea circulation and a accurate evaluation of the water masses exchange through the Sicily and the Sardinia Channels. Numerical model simulations constitute an important tool to study and to contribute to a better understanding of the multi-scale circulation and of its time variability, including those areas characterized by lacking of observational data or scarcity of appropriate time series [1].

In the past, many numerical models have been used to investigate the long-term time variability at basin and sub-basin scale [16, 3, 17, 18] or through processes-oriented studies [19, 20, 21]. On the other hand, high resolution models with horizontal numerical mesh sizes below 5 km, able to simulate the mesoscale features and its variability, have been poorly used mainly due to computational constraints. This poses the need to nest a hierarchy of successively embedded model domains, the Limited Area Model, for the downscaling of the large hydrodynamics basin scale from the coarse resolution model to finer grids in shelf and coastal areas through the nesting technique [22, 23, 24]. The use of the Limited Area Model, at regional scale embedded into a coarse resolution model, represents an efficient solution to downscale the model solutions from the basin-scale (~12.5 km) to the regional scale (~3.5 km) through a one-way off-line nesting at the lateral open boundaries. This method was found to be computationally efficient and sufficiently robust to transmit information across the connecting boundaries without excessive distortion [23, 25].

One aim of this work is to study the temporal and spatial variability of the surface water masses properties and the sea circulation in the Sicily Channel and surrounding areas. The level of mesoscale resolution achieved by our Limited Area Model allows the spatial and temporal evolution of the changing flow patterns, triggered by internal dynamics, to be followed in detail and compared with remotely-sensed and in-situ hydrological observations. Such a fine structures cannot be resolved by coarser models. Moreover, the model domain extends down to the African coast, involving the Tunisian and large part of the Libyan continental shelves areas thus giving a valuable contribution to the knowledge of the variability of this Mediterranean area.

The high resolution model has been forced at the surface with atmospheric data obtained from the ECMWF 6-hour analysis in order to reproduce the time-variability of its dynamics, while lateral forcing comes from the daily mean analysis fields of the Ocean General Circulation Model (MFS831) covering the whole Mediterranean basin [22].

In the next section the spatial and temporal scale variabilities of the hydrology and surface circulation are briefly reviewed, while the paragraph Methods in Section 3 describes the characteristics of the used model including the nesting method and the atmospheric coupling. The section follows with the presentation of the remotely-sensed, in-situ hydrological observations and the Empirical Orthogonal Function (EOF) method, while the model validation is described in Sect. 4. The space and time variability derived from the results of the model with the evaluation against in-situ observation are shown in Sect. 5 followed by the conclusions presented in Sect. 6.

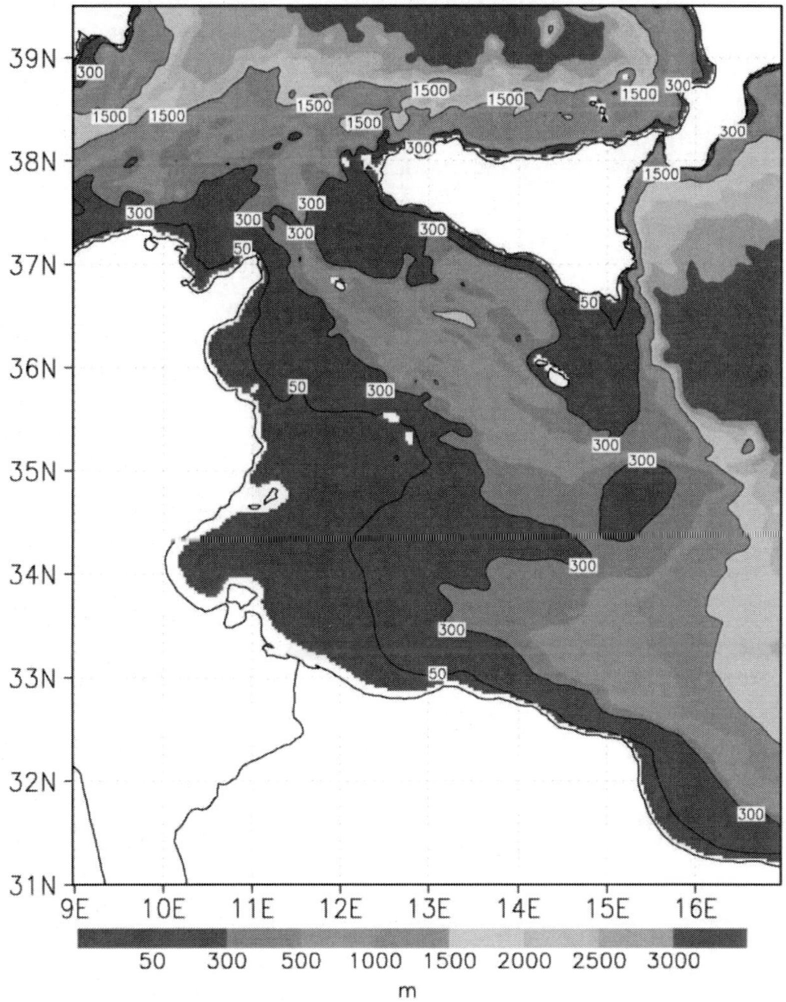

Figure 1. The SCRM domain and bathymetry interpolated into the model grid. Depths are in meters.

1. The Sicily Channel Hydrography and Surface Circulation: An Overview of Recent Progress

1.1. The Area

The modelled area (Figure 1) covers the central Mediterranean region, including the Sicily and the Sardinia Channels and the southern Tyrrhenian Sea, characterized by a rather complex topography of the sea bottom. The Sicily Channel is an intermediate area between the western and the eastern Mediterranean sub-basins (WMED and EMED, respectively). It shows shallow banks along the Tunisian and Sicilian coasts and deep trenches with flat bottom between water depths of 1100 m and 1650 m and situated in its central part. The Tunisia-Sicily transect, between Cape Bon (Tunisia) and Cape Lilibeo (Sicily), provides the direct interface between EMED and WMED, since it constitutes the main exchange area for the superficial and intermediate water masses between the two sub-basins. The Tunisian and the Libyan continental shelves are very wide and cover more than one-third of the aerial extent of the Sicily Channel. In the Gulf of Gabes the bathymetry is shallower than 30 m for large stretches away from the coast.

The Sardinia Channel is a wide, triangular-shaped region between Tunisia-Sardinia-Sicily characterized by a narrow trench, NE_SW trending, having a maximum depth of about 1900 m and wide shallow banks along the northern Tunisian coast. This channel allows for exchanges up to the deep waters between the WMED and the Tyrrhenian Sea and constitutes a strong physical boundary for the water masses flowing from or to the WMED, such as the Atlantic Water (AW), the Winter Intermediate Water (WIW), the Levantine Intermediate Water (LIW) and the deep water masses [1, 26, 27, 28].

Finally, the southern Tyrrhenian Sea shows a very restricted northern Sicilian shelf, with a steep slope, reaching a flat abyssal plain at water depths exceeding 3000 m. The Tyrrhenian Sea plays an important role in strongly modifying some of the water masses that contribute to the large scale basin circulation, and in particular the intermediate and deep waters masses.

1.2. The Hydrography and Surface Circulation

Traditionally, the hydrography of the study area can be schematized as a multi-layer system, from the surface to the bottom [7], composed of a series of adjacent water masses, both horizontally and vertically [1].

The upper layers (0-100 m) are occupied by relatively fresh water of Atlantic origin subject to a strong seasonal and mesoscale variability, which is a permanent characteristic of the region. This superficial fresh water is frequently AW (Atlantic Water) coming from the Algerian Basin. Its core depth varies seasonally and is characterized by a subsurface salinity minimum [29, 15]. The AC (Atlantic Current) enters the Sardinia Channel as an unstable coastal boundary current subject to a significant mesoscale variability and a quite complex surface pattern due to the bottom topography [10, 1, 12]. It flows eastwards along the Tunisian coast, following approximately the 200 m isobath as a broad homogeneous layer, whose characteristics change with time, especially in terms of width, core depth and volume transport [15]. In the Algerian basin, as in the whole central Mediterranean region, the lateral variation in density between the AW and the surrounding waters, together with strong differences in the velocity shear, renders the AC highly unstable with the generation of rather energetic eddies that propagate at a few km/day [10]. Part of the AW inflow re-circulates in the Sardinia Channel and returns directly into the WMED along the southern coast of Sardinia constituting the East Sardinia Current [30, 27]. The other part of the AW, Finally, the southern Tyrrhenian Sea shows a very restricted northern Sicilian shelf, with a steep slope, reaching a flat abyssal plain at water depths exceeding 3000 m. [31, 32] and of a combined effect of topography and LIW flow [27]. The first branch directly flows into the Tyrrhenian Sea along the northern coast of Sicily and forms the Tyrrhenian Bifurcation Current (TBC). It carries out 1/3 of the incoming transport [21] and spreads into the Tyrrhenian Sea with a high seasonal variability, especially in mean and total velocity [25, 18]. The remaining 2/3 of incoming transport is carried out by a second branch, turning southwards toward the Sicily Channel as a strong, narrow jet in correspondence of the deepest part of the Sicily Channel [19]. Here the flow enters into the Sicily Channel flowing eastward in two-jet structures spanning the upper 100 m with an associated minimum of salinity: the Atlantic-Tunisian Current (ATC) flowing over the Tunisian continental slope [32], and the Atlantic Ionian Stream (AIS) along the southern coast of Sicily, as inferred by hydrographic data [7, 33, 34]. This bifurcation pattern is slightly different from what estimated with in situ data by [11, 1, 12, 35]. They demonstrated that the AW flow directly splits in three veins at the entrance of the Sicily Channel following separate tracks with the minimum of salinity close to Tunisia. On the contrary, during the AIS investigation [7], the ATC/AIS bifurcation has not been observed but only the AIS with southern branches in the Sicily Channel and Ionian Sea in a complicated meandering path. The ATC is a flow into the Sicily Channel characterized by a minimum of salinity and a significant long-term variability, in space and time, as observed from surface drifters [36], Sea Surface Temperature (SST) satellite images [37] and numerical modelling results [23, 32]. The ATC flows eastward close to the Tunisian and Libyan continental shelf breaks just off Lampedusa or confined into the central part of the Sicily Channel as an energetic stream with a significant ageostrophic component [35]. The northern branch of the AW into the Sicily Channel constitutes the energetic and meandering AIS, moving eastwards along the southern coast of Sicily as a typical waveform with well defined troughs and crests, associated with cyclonic and anticyclonic structures. The AIS is a summer free jet with complex space and time variable patterns [20]. It appears to be a baroclinic wave, trapped by topography and coastal geometry [32]. A summer feature is the upwelling located along the westward and southern coast of Sicily induced by westerly winds and by inertia of the isopycnal domes of the AIS meanders and cyclonic vortices [7]. Due to the presence of large mesoscale phenomena in the Sicily Channel, the upwelling can spread far offshore, as documented by infrared satellite observation [37, 38].

2. Methods

2.1. A Brief Introduction on the Ocean Modeling

The comprehension of ocean dynamics is of fundamental importance to forecast the state of the sea. Ocean can't be separated from the atmosphere, with which there is a strong interaction in terms of exchange of energy. So water motion is driven by sun energy and Earth rotation. The most obvious way in which the Sun drives the oceanic circulation is through the atmosphere circulation, it's to say through winds but energy is transferred from winds to the ocean upper layers through frictional force. Moreover, sun also drives ocean circulation by causing variations in temperature and salinity of seawater, which turn in control its density. Changes in temperature are caused by heat fluxes across the air-sea boundary; changes in salinity are due to addiction or removal of freshwater (evaporation, precipitation, rivers waters, melting ice). All these processes are linked directly or indirectly to solar radiation. Instead, Earth rotation contributes to ocean circulation through the Coriolis force as the earth rotation causes deflection of currents and winds, irrespective to their initial direction and depending on the latitude (is zero to the Equator and maximum at the poles). This force acts to the right in the northern hemisphere and to the left in the southern one. All phenomena correlated to the Ocean dynamics can be described by a set of physical equations which takes into account seven variables: velocity **u** (three components), temperature T, salinity S, density ρ and pressure p.

- *Newton's law of motion* written for a fluid continuum

$$\rho \frac{du}{dt} = \nabla p + \rho \nabla \phi + \mathrm{A}(u) \tag{1}$$

The above equation express that the product between density (mass per unit volume) multiplied the acceleration is equal to the sum of the pressure gradient force, the body force and the force A, the frictional force in the fluid; the equation can be divided into three components:

$$\frac{\partial u}{\partial t} + u\frac{\partial u}{\partial x} + v\frac{\partial u}{\partial y} + w\frac{\partial u}{\partial z} - fv = -\frac{1}{\rho} + A_H\left(\frac{\partial^2 u}{\partial x^2} + \frac{\partial^2 u}{\partial y^2}\right) + A_V\frac{\partial^2 u}{\partial z^2} + \nu\nabla^2 u$$

$$\frac{\partial v}{\partial t} + u\frac{\partial v}{\partial x} + v\frac{\partial v}{\partial y} + w\frac{\partial v}{\partial z} + fv = -\frac{1}{\rho} + A_H\left(\frac{\partial^2 v}{\partial x^2} + \frac{\partial^2 v}{\partial y^2}\right) + A_V\frac{\partial^2 v}{\partial z^2} + \nu\nabla^2 v \tag{2}$$

$$\frac{\partial w}{\partial t} + u\frac{\partial w}{\partial x} + v\frac{\partial w}{\partial y} + w\frac{\partial w}{\partial z} = -\frac{1}{\rho} + A_H\left(\frac{\partial^2 w}{\partial x^2} + \frac{\partial^2 w}{\partial y^2}\right) + A_V\frac{\partial^2 w}{\partial z^2} + \nu\nabla^2 w$$

where A_H and A_V are called the horizontal and vertical turbulent viscosity coefficients respectively [39].

- *Continuity equation*

$$\frac{1}{\rho}\frac{\partial \rho}{\partial t} = -\nabla u \tag{3}$$

the ratio between the temporal total variation of density ρ and the density itself is equal to the opposite of the velocity field divergence.

- *State equation*

$$\rho = \rho(S, T, p) \tag{4}$$

The aim of ocean modeling is to use ocean physic lows to obtain a forecast of the sea state. This could be done through the help of computer calculation, for which is impossible to implement all physic equations in the aforementioned form above; in fact, a number of methods that discretize such an equations exist, to make computers able to solve them and to forecast variables values. These methods can be divided into two macro categories: *finite differences methods* and *finite elements methods.*

The finite difference method (FDM) consists in approximating the differential operator by replacing the derivatives in the equations using differential quotients. Imagine to superimpose a regular grid on ocean surface and along its vertical in the way in which the domain is partitioned in space and in time; solutions approximation is computed at the space or time points, which correspond to grid points. The error between the numerical solution and the exact solution is determined by the error that is committed by going from a differential operator to a difference operator. This error is called the discretization error or truncation error. The finite element method (FEM) is nowadays applied from industrial applications to fluid mechanics. FEM has got the same aim of the FDM but its grid is irregular and usually constituted by a set of consecutive triangles, becoming smaller close to the shoreline (the smaller the triangles, the higher the number of grid points and consequently the resolution) [40]. Different calculations techniques were developed for each method to obtain the best forecast of the sea state. These techniques are the bases on which ocean models are built. There is also a variety of models that differ in spatial and temporal resolution and in the domain extension. For example, the Mediterranean Sea is covered by the Mediterranean Forecasting System - MFS and its following versions [22] which produces daily and monthly means forecasts of ocean variables. MFS's outputs constitute the initial and boundary conditions - together with surface initial conditions given by atmospheric models, like the European Centre for Medium Range Weather Forecast (ECMWF) - for models at higher resolution and lower domain extension. For this reason MFS is frequently called "father model." Following this methodology, regional models are more able to reproduce small-scale dynamics, depending on their spatial resolution (i.e., on the distance between grid points in which variables values are computed; [41]). The regional model used for the present study is the Sicily Channel Regional Model (SCRM), whose characteristics are described in the following section.

2.2. Model Description

The Sicily Channel Regional Model (SCRM) is a free surface three-dimensional primitive equation finite difference hydrodynamic model based on the Princeton Ocean Model [42]. It solves the equations of continuity, motion, conservation of temperature and salinity, assuming that the fluid is hydrostatic and that the Boussinesq approximation is valid. The density is calculated by an adaptation of the UNESCO equation of state revised by [43]. The vertical mixing coefficients for momentum and tracers are calculated using the [44] turbulence closure scheme, while the horizontal diffusion terms (eddy viscosity coefficient) are provided by the Smagorinsky parameterization [45].

The SCRM has been implemented in the region defined between 9°E and 17.10°E and from 31.50°N to 39.50°N with a horizontal resolution of 1/32° (~3.5 km). In the vertical it uses 24 sigma levels, denser near the surface through a logarithmic distribution. The external time step is set to 4 s with an internal one every 120 s. The model bathymetry has been obtained from the U.S. Navy Digital Bathymetric Data Base-DBDB1 at 1/60° and mapped with bilinear interpolation into the model grid. The minimum depth has been set to 5 m. Additional smoothing is applied to reduce the sigma coordinate pressure gradient error [46].

The numerical simulation of the study area has been carried out using the SCRM embedded into the MFS831. This permits to produce a more detailed description of the circulation in the region, including some mesoscale components that cannot be resolved by coarser models.

The model has been initialized at 1 January 2001 using dynamically balanced fields from MFS831 prepared through an innovative tool based on the Variational Initialization and Forcing Platform [47]. This method is largely used in ocean forecast systems [48] and drastically reduces the amplitude of the numerical transient processes, generally following the initialization phases for several days after the initialization [49].

2.2.1. Lateral Open Boundary Conditions

Lateral open boundary conditions are defined through a simple off-line one way nesting technique used to simulate a high-resolution domain embedded into a coarse resolution model. This technique represents an efficient way to downscale the model solutions from the basin scale (~12.5 km) to the regional scale (~3 km). The method has been largely used in numerical weather predictions and recently in numerical oceanography to simulate the hydrodynamics of limited shelf and coastal areas [25, 50, 23, 51]. The SCRM is nested at the lateral open boundaries with the MFS831 covering the whole Mediterranean Sea with a horizontal resolution of 1/8° [22]. The values of temperature, salinity, total velocity and elevation are transferred from the coarse spaced grid of MFS831 to the finely spaced grid of the SCRM open boundaries through an off-line, one-way asynchronous nesting. The coupling between MFS831 and SCRM is realized by imposing the interpolation constraint on the total velocity. This allows the total volume transport to be preserved after the interpolation procedures from the coarse to the fine resolution model. The vertical integrated velocity component normal to the open boundaries at each time step is specified following the method developed by [22];

$$\bar{U}_{high} = \bar{U}^{\text{int}}_{coarse} \frac{H}{(H + \eta_{high})} + \varepsilon \sqrt{\frac{g}{(H + \eta_{high})}} \eta_{high} \tag{5}$$

where

$$\overline{U}^{\text{int}}_{coarse} = \frac{1}{H}\int_{-H}^{\eta} U_{coarse}dz$$

is the MFS831 vertically integrated velocity component normal to the boundaries interpolated into the SCRM. In cases of inflow through the open boundaries, the three-dimensional temperature and salinity fields are prescribed from the values of the MFS831 solution interpolated on the SCRM open boundaries:

$$(\theta,S)_{high} = (\theta,S)^{\text{int}}_{coarse}.$$

The free surface elevation is not nested (zero gradient boundary condition), then:

$$\frac{\partial \eta_{high}}{\partial \vec{n}} = 0.$$

2.2.2. Surface Boundary Conditions

At the surface, the SCRM is driven by an interactive air-sea module. This consists in the use of a well-tuned set of bulk formulae for the computation of momentum, heat and freshwater fluxes at the air-sea interface, where fluxes depend upon the state of the ocean directly and by the 6-hours (00:00, 06:00, 12:00, 18:00 UTC) ECMWF analysis fields of the atmospheric parameters (wind at 10 m a.s.l., temperature at 2 m a.s.l., cloud cover, specific humidity and pressure). Then momentum, heat and freshwater fluxes are mapped on the SCRM grid through bilinear interpolation, and then linearly interpolated in time for each model time step.

The surface boundary condition for the momentum flux (wind stress) is derived from the wind components by using a constant drag coefficient CD = CD(TA,TS,W) as a function of the wind amplitude (W), the air (TA) and the SSTs (TS) through the polynomial approximation given by [52]; TS data are directly taken from the ocean model.

The surface boundary condition for potential temperature takes the classical form:

$$K_H \left.\frac{\partial \theta}{\partial z}\right|_{z=\eta} = \frac{Q_T}{\rho_0 C_p}, \qquad (6)$$

where KH is the vertical heat diffusivity for water, Cp (4186 J Kg-1 K-1) is the specific heat capacity of pure water at constant pressure and QT is the net heat flux or total heat budget.

The net heat flux, or total heat budget, involves the balance between surface solar radiation (QS), the net long-wave radiation (QB), the latent and sensible heat fluxes (QE and QH). We consider that the fluxes QB, QE and QH are positive for energy gained by the atmosphere. The heat flux components are calculated using the Reed formula [53] for the short wave radiation flux (QS) and the Bignami formula [55] for long wave radiation (QB). The latent (QE) and sensible (QS) heat fluxes are given by the bulk aerodynamic formulas using the Kondo scheme for the turbulent exchange coefficients [56]. For the salinity flux we consider the water balance:

$$K_H \frac{\partial S}{\partial z}\bigg|_{z=\eta} = (E - P - R)S_{z=\eta} + C_2(S_{z=0} - S_{z=\eta}) \tag{7}$$

where E = QE/LE is the evaporation rate calculated interactively, P is the monthly precipitation rate obtained from [57] and R is the river runoff. In our simulation we have imposed R = 0, due to the absence of significant rivers supply. The second term of Eq. 7 is the flux correction term used to impose a forcing to produce a sea surface salinity consistent with the seasonal climatology. The term S_z = .0 is the monthly mean climatological surface salinity from MED6 and $S_z = \eta$ is the first level of salinity predicted by the model. In this study, the value for C2 has been chosen to be equal to 0.7 m day-1.

2.3. Observational Data

A circulation model needs to be assessed in its ability to transfer results from the numerical to the physical domain. This has been done by comparing model results with in-situ and remote-sensed hydrographical observations, using different statistical diagnostics.

2.3.1. The In-Situ Data

The in-situ data come from the four oceanographic surveys named MedGOOS5, 6, 7 and 9 carried out onboard the R/V URANIA of CNR. The surveys span different seasons between 2002 and 2004 (Table 1) along a transect across the Sicily Channel with a repetition of 11-12 casts (Figure 2). At all the hydrological stations, data are acquired by a CTD rosette system consisting of a SBE911 plus mounted on a General Oceanics rosette with 24 12-litre Niskin bottles. The vertical profiles of all parameters are obtained by sampling the signals at 24 Hz, with the CTD/rosette system sinking at a speed of 1 m s-1. In-situ temperature and conductivity data are acquired by an SBE-3/F thermometer with a resolution of 10-3 °C, and by an SBE-4 sensor with a resolution of 3 x 10-4 S.m-1. The calibration for salinity is accomplished by collecting water samples at fix depths followed by on board analysis using a Guildline Autosal salinometer. The data are pre-processed on board and the coarse errors are corrected. A further processing of the data has been carried out by using the Ocean Data View software [58].

2.3.2. The Satellite Data

The pathfinder SST data derived from the 5-channel Advanced Very High Resolution Radiometers (AVHRR) on board the NOAA -7, -9, -11, -14, -16 and -17 polar orbiting satellites have been used to assess the model skill in reproducing the sea surface temperature. The monthly averaged data for the ascending pass (daytime) on equal-angle grids of 8192 pixels/360 degrees (nominally referred to as the 4 km resolution) are used. The data have been acquired from the Physical Oceanography Distributed Active Archive Center at the NASA Jet Propulsion Laboratory[1].

[1] Website: http://podaac.jpl.nasa.gov.

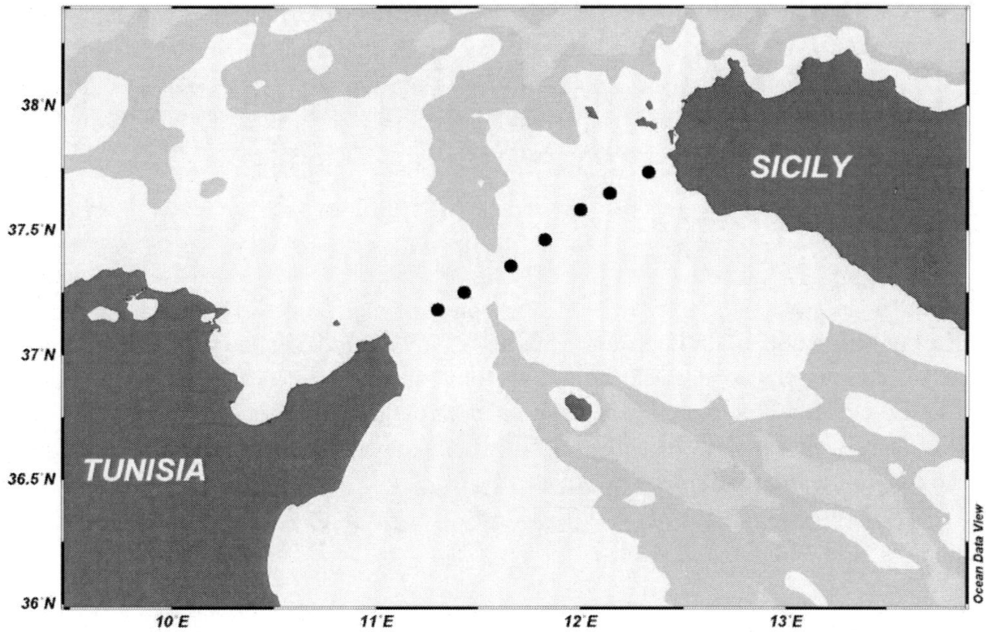

Figure 2. Horizontal distribution of some CTD casts during the survey MedGOOS9 as representative of all the casts along the Sicily-Tunisia section.

Table 1. The synoptic survey MedGOOS conducted in the Sicily Strait, the section that connects Cape Bon (Tunisia) to Mazara del Vallo (Sicily, Italy)

Cruise	Period
Medgoos 5	6 - 7 November 2002
Medgoos 6	31 March - 1 April 2003
Medgoos 7	11 - 12 January 2004
Medgoos 9	6 - 7 October 2004

2.4. Empirical Orthogonal Functions (EOFs) Decomposition

Qualitative analysis of oceanographic features and basic statistics of model outputs have been further supported by EOF analysis. The EOF method, introduced in meteorology by [56], has been widely used in oceanography [60, 61, 62]. EOFs aim is to obtain relevant informations from a complex dataset, removing redundancy, which is represented by the autocorrelation. This kind of analysis is also called "Principal Components Analysis (PCA)" which geometrically has got the goal to represent data in a reference plan that mainly highlights EOFs structure, extracting important patterns from measurements of ocean (and atmospheric) data. Specifically, EOFs give the possibility to reduce the dimension of large datasets to few orthogonal (uncorrelated) modes of variability [63]. Usually, most of the variance of a spatially distributed series is in the first few orthogonal modes whose patterns may be linked with physic features of the phenomenon under study. However, there is no direct physical or mathematical relationship existing between the statistical EOFs and any related mode [64]. Generally, oceanographic data are organized in arrays containing a three dimensional field F (two dimension in space and one dimension in time) for each vertical level. So the field F is function of time t, latitude θ and longitude ϕ. Horizontal coordinates are discretized to yield θ_j, with j = 1, .., p_1, and ϕ_k, with k = 1, .., p_2 and similarly for time, t_i, with i=1, .., n. So the total number of grid point is $p = p_1\, p_2$. The field F is:

$$F_{i,j,k} = F(t_i, \theta_i, \varphi_k) \tag{9}$$

Then it is possible to transform F in a two dimensional field (X) by ordering the single spatial arrays (function of latitude and longitude) in ordered vectors. In this way we can imagine to have a gridded data set composed of a space-time field *X(t,s)* representing the value of the field X at time *t* and space position *s*. At the discrete time t_i and at the position s_j, the field value is x_{ij}, for i=1,.. n and j=1,.. p, where remember that $p=p_1p_2$. At each jth grid point we can define the field time average as

$$\overline{x_j} = \frac{1}{n}\sum_{k=1}^{n} x_{kj} \tag{10}$$

The anomaly field at (t, s) is defined as

$$x'_{ts} = x_{ts} - \overline{x_s} \tag{11}$$

or in matrix form:

$$X' = X - A\overline{x} \tag{12}$$

where $A=(1, ...1)^T$

Once the anomaly data matrix above (12) is determined, the covariance matrix is then defined as

$$C = \frac{1}{n-1} X'^T X' \tag{13}$$

which include the covariance (a measure of the strength of the correlation between two or more sets of random variables) between any couple of grid points.

The aim of the EOFs (or PC) is to find grid points that explain maximum variance, that is to find a direction $a = (a_1,...a_p)^T$ such that *X'a* has the maximum variability. To make the problem bounded, it's required that the vector a is unitary, hence the problem is traced to an eigenvalue problem, that is:

$$Ca = \lambda a \tag{14}$$

By definition, the covariance matrix C is symmetrical and diagonalizable. The kth EOF is simply the kth eigenvector a_k of C after the eigenvalues and the correspondent eigenvectors

have been sorted in a decreasing order. Because C is diagonalizable, the set of eigenvectors form an orthogonal basis of the p-dimensional Euclidean space and so, by construction the EOFs are orthogonal and the PCs are uncorrelated. The orthogonality property provides a complete basis where the time-varying field can be separated as

$$X'(t,s) = \sum_{k=1}^{M} c_k(t) a_k(s) \quad (15)$$

However, this property puts a limit in the physical interpretability of each EOF since tends to be non orthogonal [65] in general physical patterns. A truncation order can be applied to equation (15), obtained by fixing the amount of variance, e.g., 80%, and it is possible to choose the leading EOFs that explain altogether this amount of variance.

The EOF decomposition method is applied on the dataset composed by the monthly mean fields coming from the SCRM during the simulated period from January 2001 to December 2004. The modes of variability, also known as EOFs (EOF-1, EOF-2, ..., EOF-n), contain information on the dataset variability. The interpretation of the EOF modes just aims to relate the data modes with physics. The temporal amplitudes associated with the spatial patterns, referred to as EOF expansion coefficients or simply Principal Components (PCs) [66], are also object of physical interpretation. A detailed description of the EOF method applied to geosciences can be found in [67].

2.5. Model Validation and Statistic: Root Mean Square Error (RMSE)

The Root Mean Square Error (RMSE) has been used as a standard statistical metric to measure model performance in oceanography, meteorology, air quality, and climate research studies as it provides a complete picture of the error distribution [68].

For the present study, RMSE gives a measure of the discrepancy between known satellite observations and the corresponding values simulated by the model. It is defined by:

$$RMS = \sqrt{\sum_{k=1}^{N} \frac{(H_{(i,j)} - O_{(i,j)})^2}{N}} \quad (16)$$

where the model error (bias) is the difference between the simulated, H(i, j), and the observed, O(i, j), values, while N is the total number of available data values.

3. Results and Discussion

3.1. Validation of Sea Surface Temperature

The comparison between the model basin averaged SST and the corresponding averages from remotely sensed observations is shown in Figure 3A which superimposes the respective time series of monthly means. There is an overall agreement for a large part of the annual cycles. A better assessment is made by means of the RMS error in the monthly averages

(Figure 3B). The error shows the model overestimation in the SST especially in summer. The model is on average 0.4°C warmer than satellite observations. The summer surface temperature overheating of the model can reach up to 1.5°C in August. Assuming that we can fully rely on the accuracy of satellite measurements, the summer overestimation could be caused by various factors. The spatial distribution in the annual average of the SST RMS error in Figure 4 for the year 2004 indicates that the relatively higher RMS error in the shallower areas of the domain, such as over the Tunisian shelf, can be attributed to the computation of the diffusion terms that may constrain the heat flux to stay close to the surface. Other reasons could be related to the inaccuracy in the physics of the bulk formulae that are used for the derivation of heat fluxes and the coarse horizontal resolution of the atmospheric forcing. The reduced numerical resolution in the vertical dimension, due to the thickening of the first sigma layer in the deeper regions of the domain, may explain the higher errors over the western Ionian Sea [24].

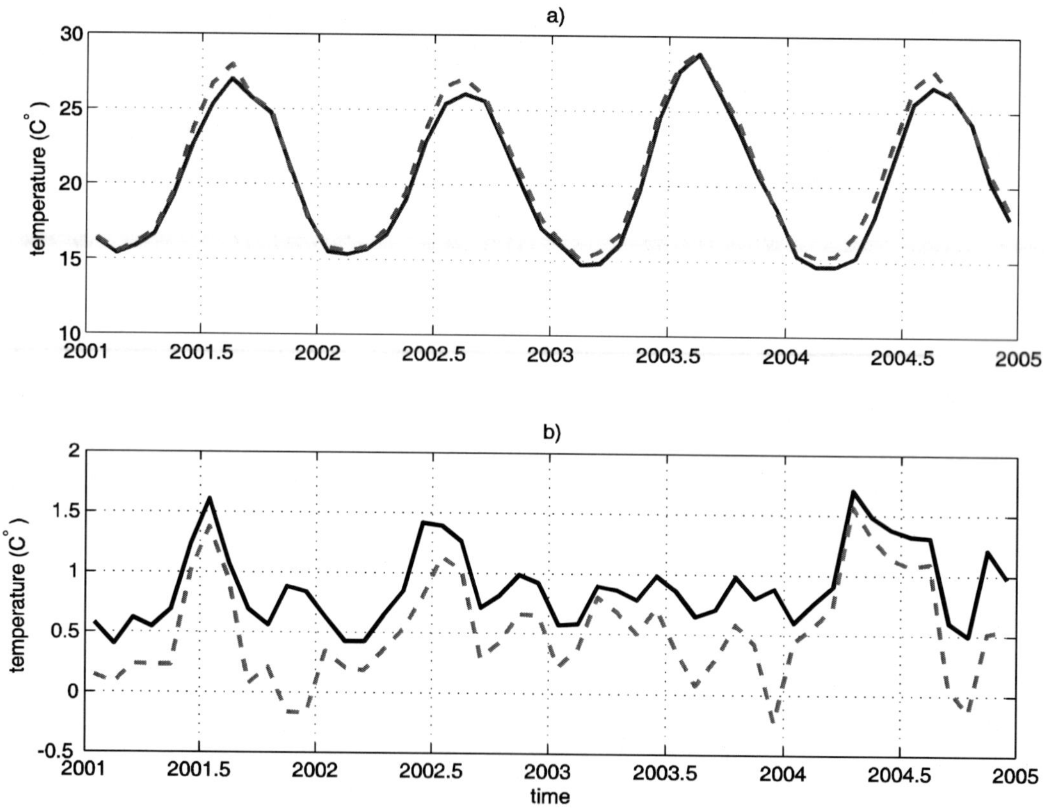

Figure 3. a):Monthly time series of the basin averaged SST during the simulated period (2001-2004). Model hindcast (dashed line) and mean of satellite SST observations (solid line). b): Time series of the basin averaged hindcast RMS error (solid line) and BIAS (dashed line) computed with respect to the observed SST.

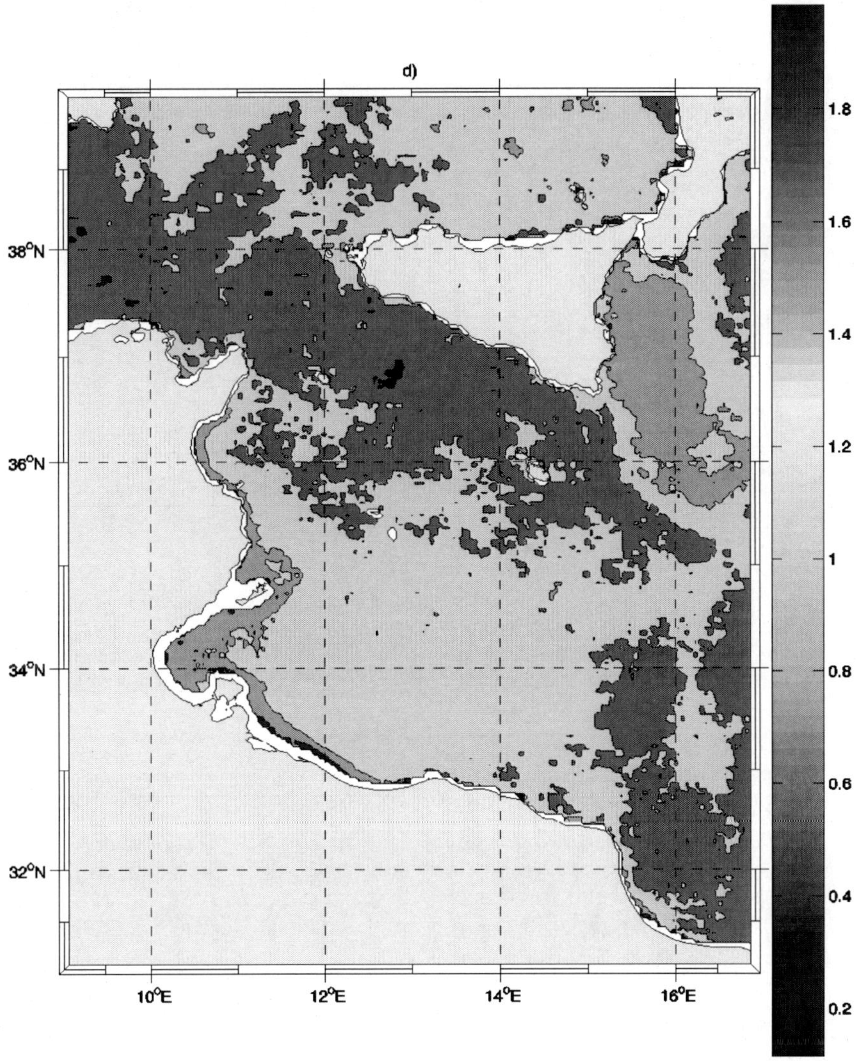

Figure 4. Example of spatial distribution of the annual averaged RMS error map between the satellite and model SST for the year 2004. Units are in °C.

3.2. Validation of the Simulated Water Column Hydrology

The CTD profiles acquired during the four MedGOOS cruises have been compared on a point by point basis with the SCRM model profiles through the computation of the RMSE of temperature and salinity. The RMS error (for temperature and salinity profiles) is presented in Figure 5 as an averaged vertical profile for each survey.

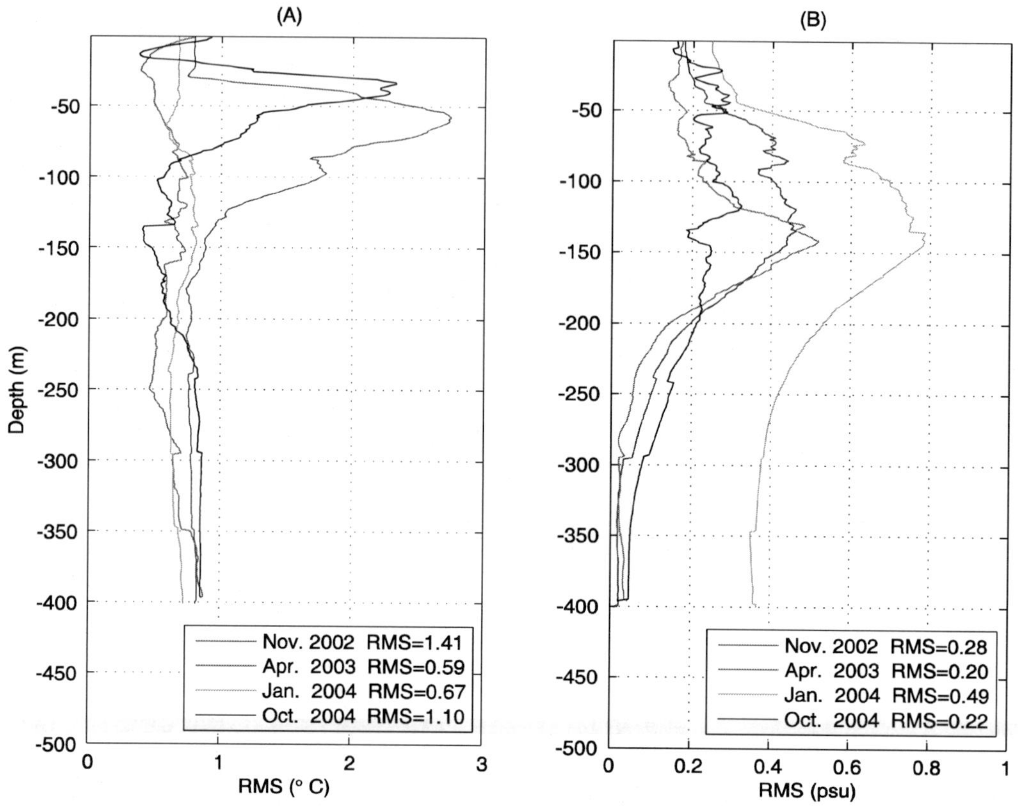

Figure 5. The cruise averaged RMS error between model temperature in (a) and salinity in (b) and CTD observations calculated at different depths (blue line – November 2005; red line - April 2003; green line - January 2004; black line - October 2004) as function of depth at the Sicily Strait for: A) temperature (°C); B) salinity. The vertically averaged RMS errors are written in the legend box of the respective panels.

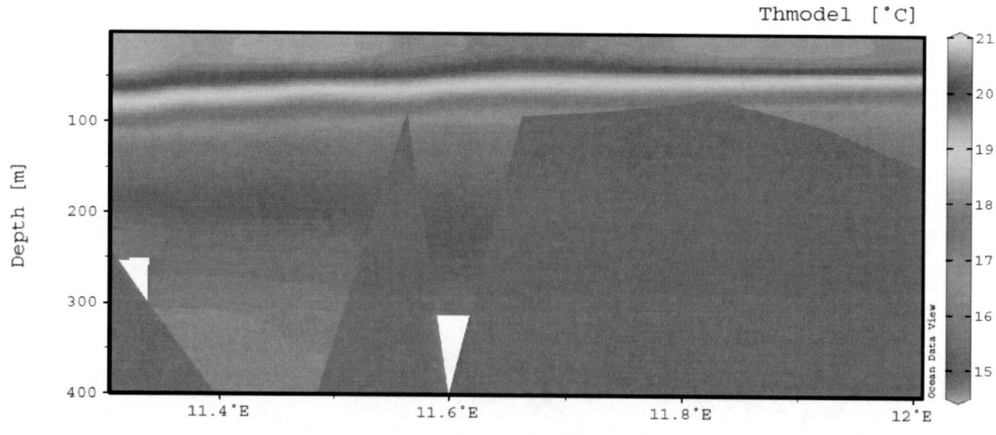

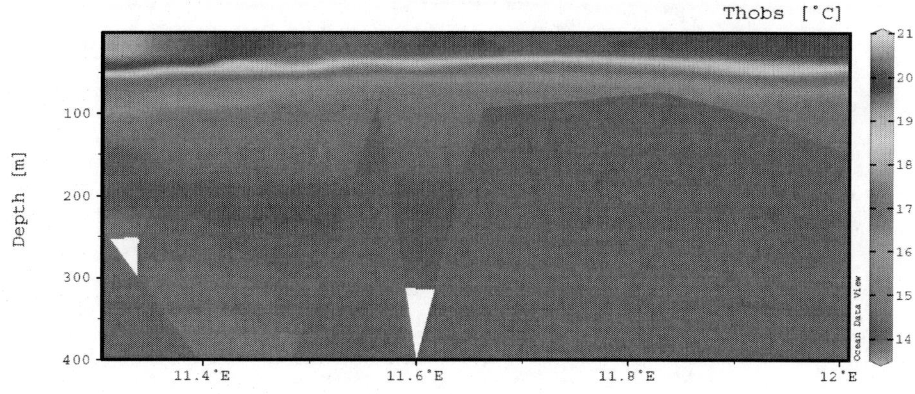

Figure 6. Comparison between the observations (upper panel) and hindcasted fields (lower panel) during the survey MedGOOS 5 (November 2002). Units are in °C.

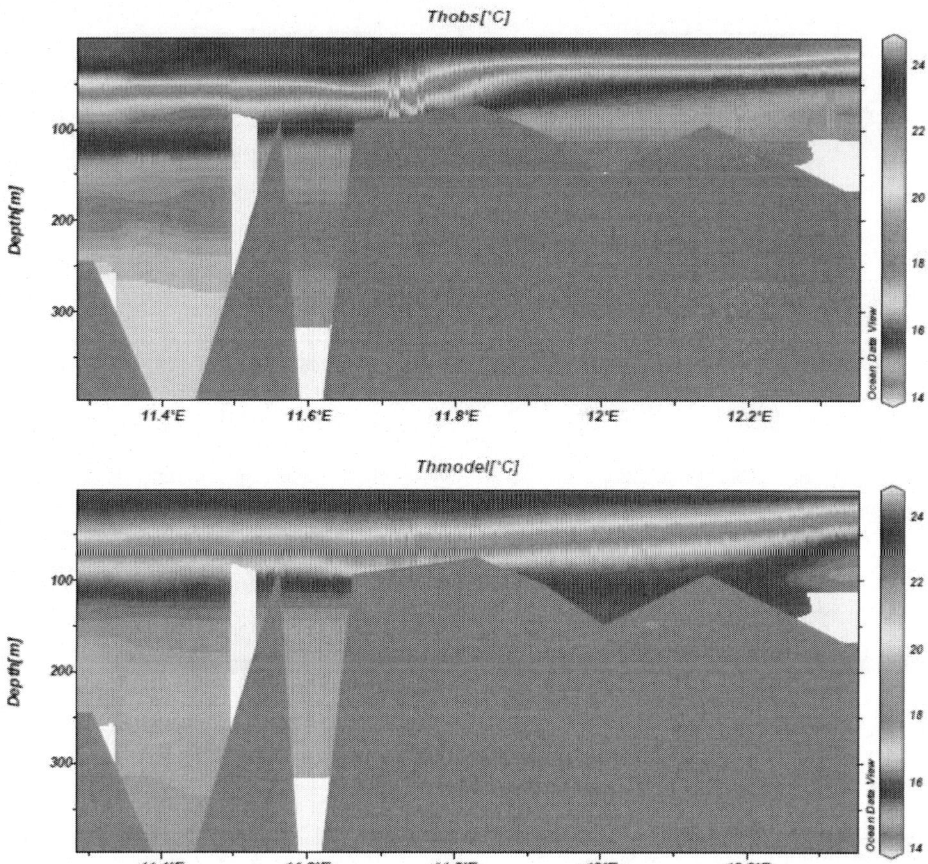

Figure 7. Comparison between the observations (upper panel) and hindcasted fields (lower panel) during the survey MedGOOS 9 (October 2004). Units are °C.

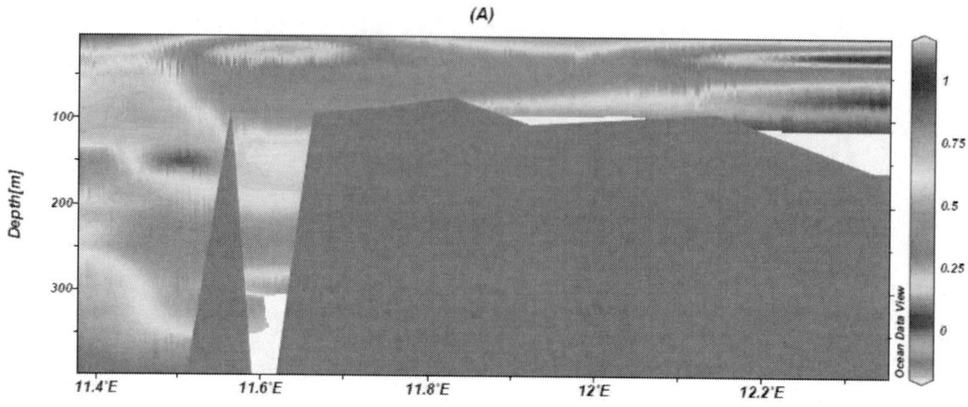

Figure 8. Bias of temperature between hindcast values and observations during the survey MedGOOS 6 (April 2003). Units are in °C.

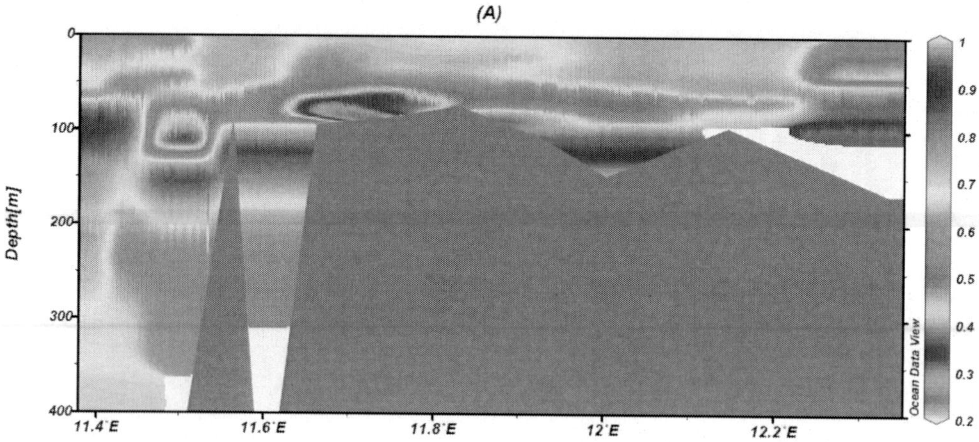

Figure 9. Bias of temperature between hindcast values and observations during the survey MedGOOS 7 (January 2004). Units are in °C.

Figure 5A shows that the RMS error between the model temperature and observations, for the spring cruises MedGOOS6 and MedGOOS7, is less than 1°C along the entire water column, while during the MedGOOS5 and MedGOOS9 (conducted in fall season) it reaches a maximum of ~2.5°C and ~2.1°C, respectively, between 40 and 80 m water depth. At the surface and at the bottom the differences do not exceed 0.8°C in all the cruises. As evidenced in Figures 6 and 7 the model tends to smooth out the profile out during periods of strong stratification, especially at the sharp gradient of the transition layer, between the surface and the intermediate water masses. This is probably related to the vertical mixing scheme used by the model, which mainly affects the upper layers. During periods of weaker stratification, such as during the surveys in April 2003 and January 2004, the bias does not exceed +1.5°C and appears located mainly below 100 m to the bottom on the Adventure Bank (Figures 8 and 9). Similarly, concerning the salinity, the mismatch in its profiles is influenced by the water column stratification, and particularly by the vertical gradients in the transitional depths between the surface and the intermediate water masses. Salinity error profiles (Figure 5) are all characterized by an increasing trend from the surface (0.2) to about 150 m, which represents the transitional layer between the surface AW and the intermediate LIW water masses. Below this depth, all vertical error profiles decrease to the bottom where the errors do not exceed 0.1, except for the observations in January 2004 (i.e., MedGOOS 7). This particular profile is characterized by a maximum error in salinity of about 0.8 at the transition layer while it is 0.25 at the surface and 0.4 at the bottom. The bias of salinity (Figure 10A-D) shows that at the surface the model always overestimates the observations, especially for the AW associated to the ATC transport, and less for the waters associated with the AIS. As an example see Figure 10. The origin of this error is believed to be mostly originating from the lateral boundary values provided by MFS1831. The maximum bias in salinity is located at the interface between AW and LIW, where the model underestimates the observations with a maximum bias of about -1 during the observation made in January 2004 (Figure 10B). The model is thus unable to reproduce the sharp vertical changes as the errors may also relate to a wrong location and estimate of the LIW core due to an excessive mixing of the LIW with the surrounding water masses in the model.

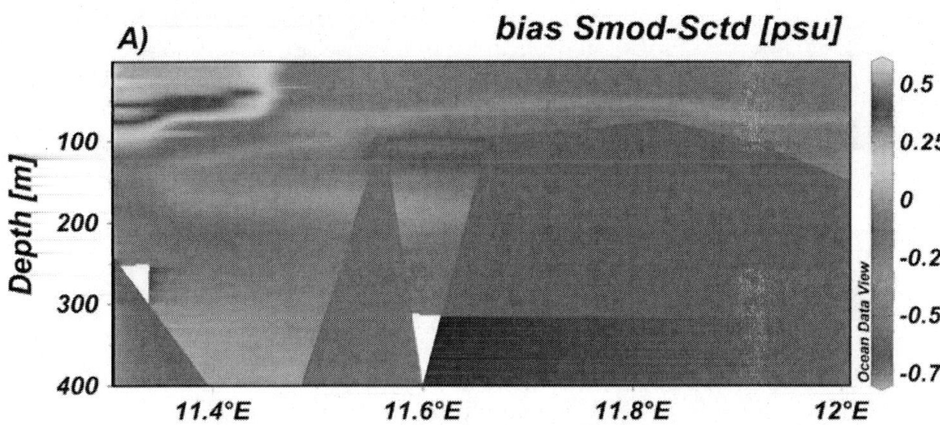

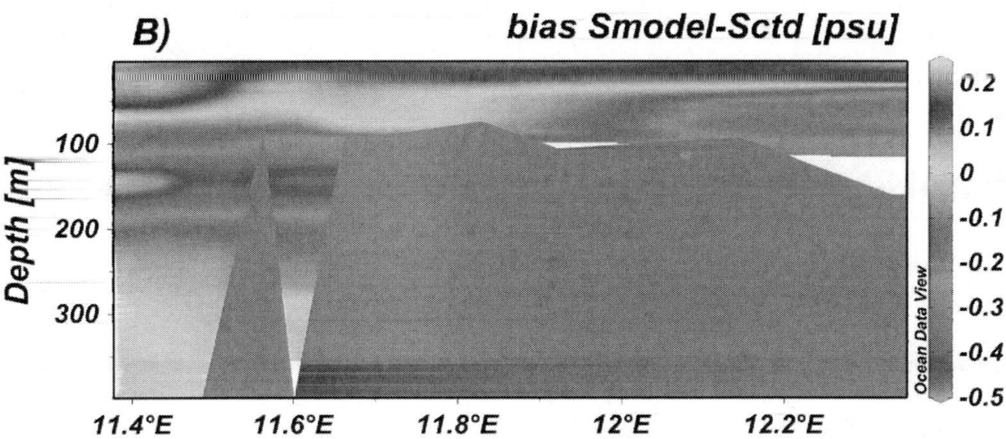

Figure 10. (Continued).

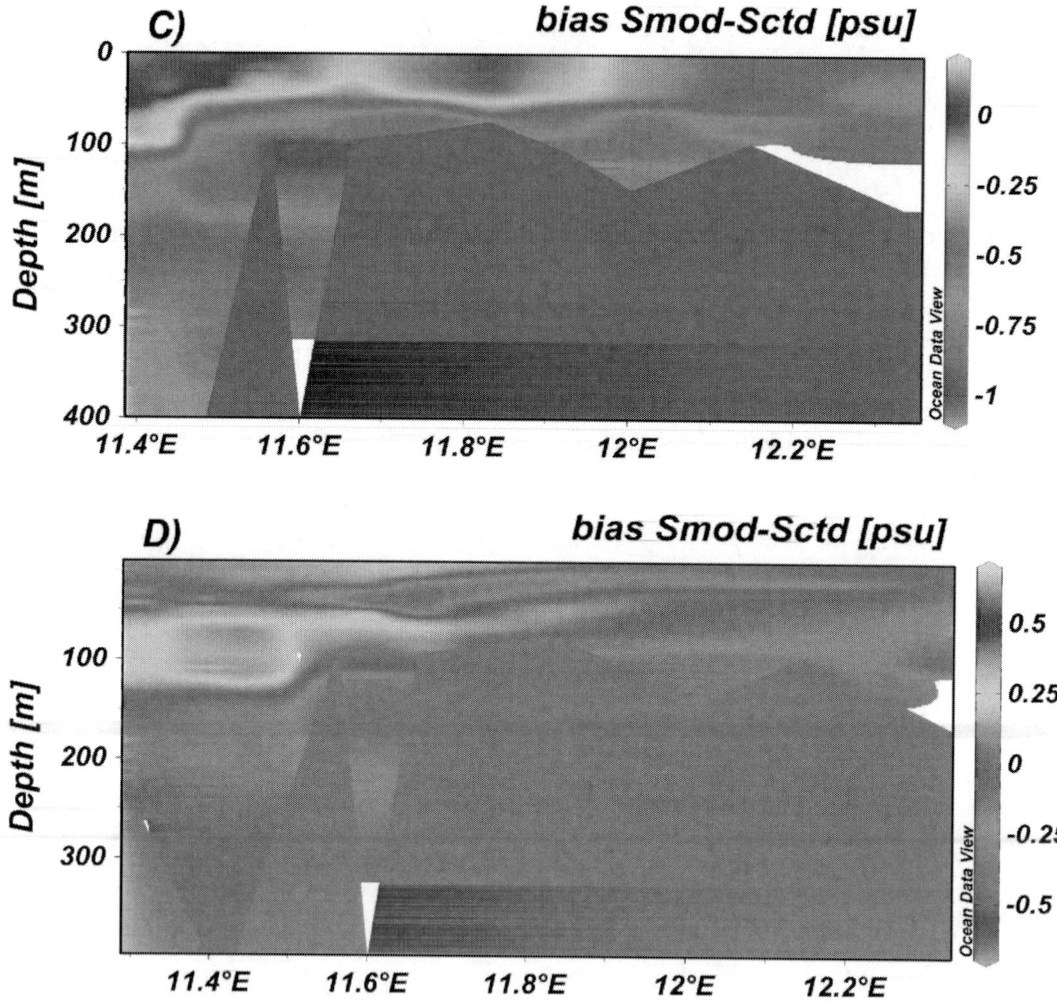

Figure 10. Bias of salinity between hindcast values and observations during the surveys in November 2002 (A), April 2003 (B), January 2004 (C) and October 2004 (D).

In the light of such an evaluation, we argue that the performance of the model in reproducing real profiles (taking into consideration the issues of comparing point station data with model gridded data) is in agreement with those recently estimated for other regional hydrodynamic applications [69, 70].

3.3. Seasonal and Interannual Variability

Within the limits of the model skill expressed in the previous section, the results of the model can be analyzed to provide an improved knowledge of the oceanographic circulation in the study area.

The daily volume averaged temperature time series in Figure 11A show a large seasonal variability characterized by a summer maximum temperature (about +14.6°C in September), which is rather stable from year to year. Winter minimum temperature (generally in March) ranges between +14.12°C (in 2001) to +14.20°C (in 2002). Time series of the surface averaged temperature (Figure 11B) shows the same characteristics with a less evident interannual signal during winter and greater in summer with an absolute temperature maximum of 28.70°C in August 2003. The annual maximum/minimum basin averaged temperature occurs about one month later than the surface ones, due to the heat diffusion times, and about three months after the corresponding extremes of the net heat flux (not shown), thus indicating the time scale for the vertical heat exchange across the water column.

The annual cycle of the daily volume basin averaged salinity (Figure 11C) shows a seasonal cycle characterized by maximum values in December, and minima reached in different periods (April 2001, May 2002, July 2003 and 2004). Moreover, it does not show a marked interannual variability, remaining approximately constant at 38.52 ± 0.07. This is a consequence of the salt flux balance at the open boundaries, namely between the inflows of fresh AW at the surface, coming from the Sardinia Channel, and the saltier LIW flowing at intermediate depth from the Medina sill, rather than to variations in the surface salt fluxes (not shown). In fact, the time series of the surface averaged salinity (Figure 11D) show similar annual cycles characterized by maximum values of about 38 in autumn and minima of 37.8 from March to May.

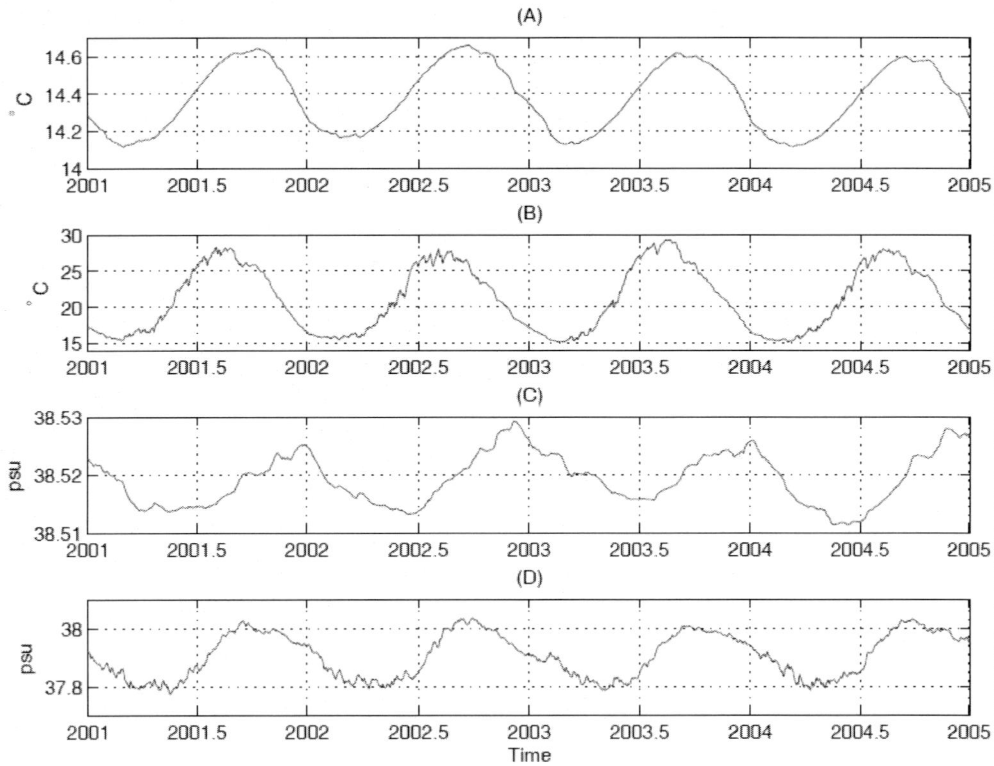

Figure 11. Daily time series of the volume and surface averaged temperature and salinity during the simulated period (2001-2004): (A) mean volume temperature; (B) mean surface temperature; (C) mean volume salinity; (D) mean surface salinity.

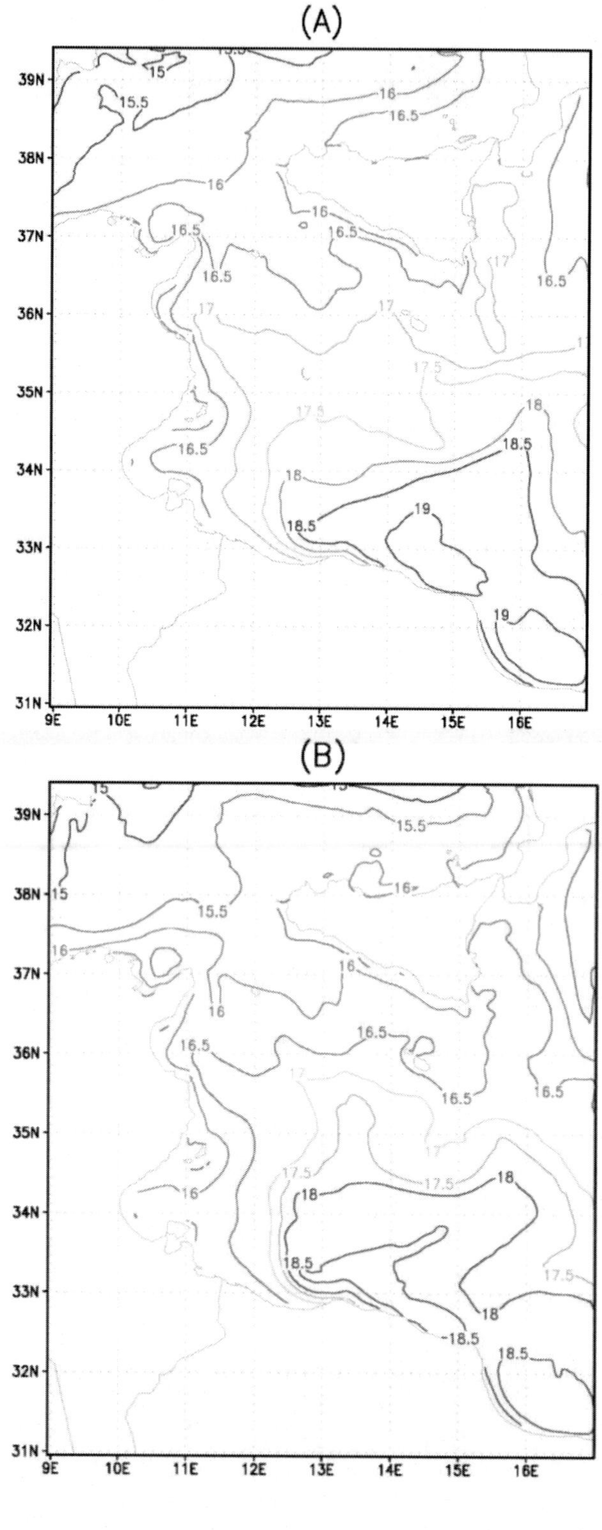
(A)
39N
38N
37N
36N
35N
34N
33N
32N
31N
9E 10E 11E 12E 13E 14E 15E 16E
(B)
39N
38N
37N
36N
35N
34N
33N
32N
31N
9E 10E 11E 12E 13E 14E 15E 16E

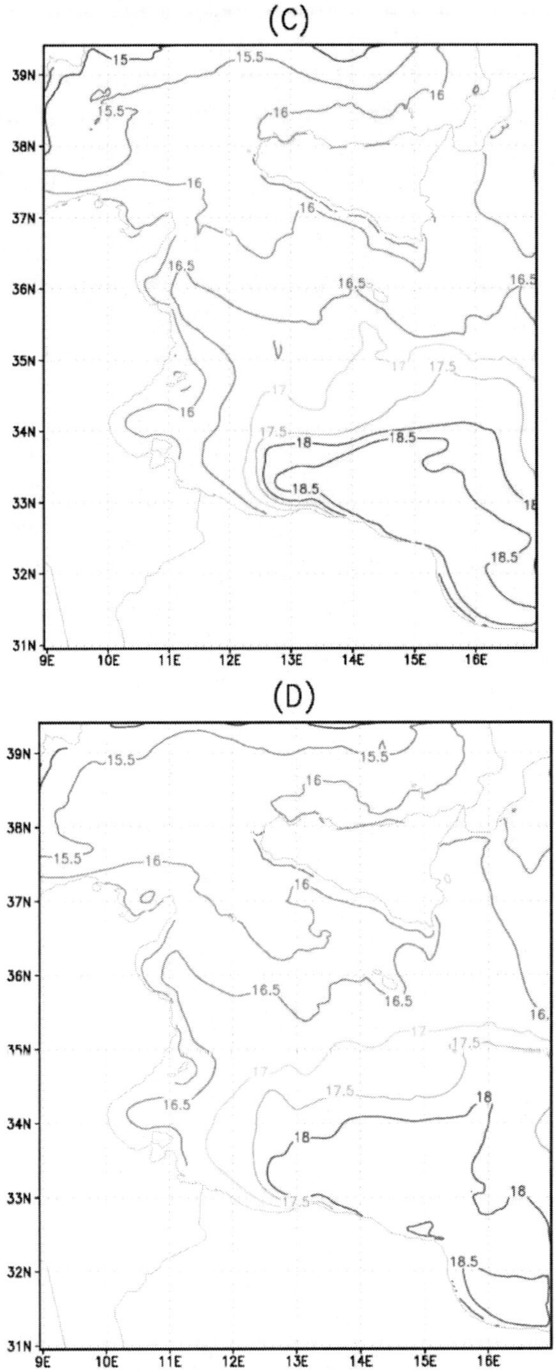

Figure 12. Near surface (5m) temperature (◦C) for winter: (A) 2001; (B) 2002; (C) 2003; (D) 2004.

The thermal structure of the SST in winter is fairly homogeneous with rather stable values from year to year (Figures 12A-D). The horizontal fields are characterized by weak horizontal gradients (positive from north-west to south-east) with large patches of relative cooler waters in the Sardinia Channel. Above 36°N the temperature does not exceed +17°C. Relatively warm waters are found at lower latitudes (on the Libyan shelf), with a maximum temperature of about +18.5°C, coinciding with large anti-cyclonic gyres. The spring temperature fields (not shown) represent a transitional period between winter and summer, with a progressive increase of the superficial temperature, in consequence of surface heating, and no significant difference can be noted throughout the years of the simulation. In summer (Figures 13A-D), the complex fields are characterized by fronts and by coastal upwelling along the southern coast of Sicily, with the southeastward advection of these cold patches forming long plumes and filaments. Warm waters are found over the Libyan and the shallow Tunisian shelves (Gulf of Gabes, with the highest values above +30°C), over the Tyrrhenian Sea, and over the eastern Ionian escarpment. The contrast in temperature of warmer Ionian waters with the westerly relative cooler waters produces a conspicuous thermal filament, known as Maltese front [25], well developed during all the years of simulation. A significant difference between the summer fields can be noted in 2003 (Figure 13C) characterized by absolute surface temperature maxima located especially over the African shelf and to the west of the Sicily Channel. These positive anomalies are due to an unusual overheating of the sea surface related to an increase of the air temperature and a sensible decrease of the wind stress [71, 72]. This combination of factors produces a sensible overheating of the surface layers, resulting in a sensible reduction of the basin and superficial averaged total kinetic energy (Figures 14A and 14B, respectively), and affecting the net heat and salt volume transport in the Sicily Channel (Figure 15) and in the Sardinia Channel (Figure 16).

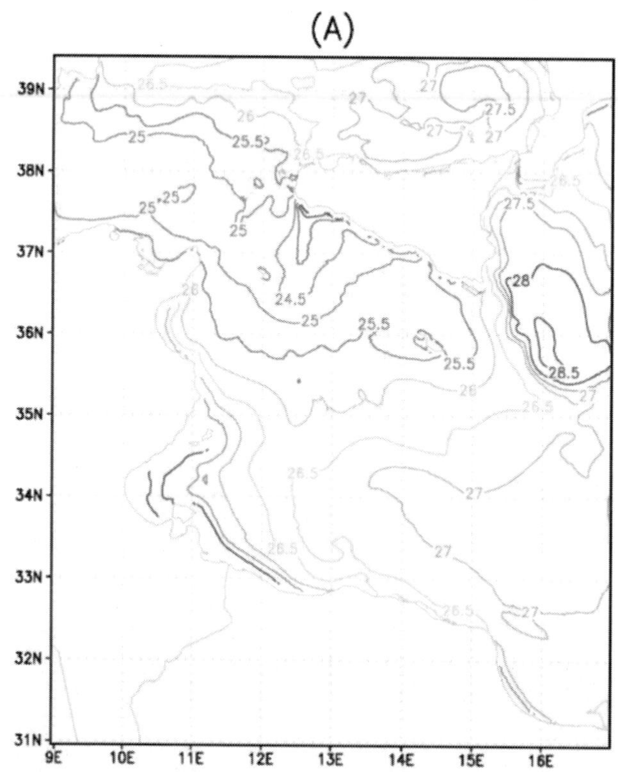

Figure 13. Near surface (5 m) temperature (◦C) for summer.

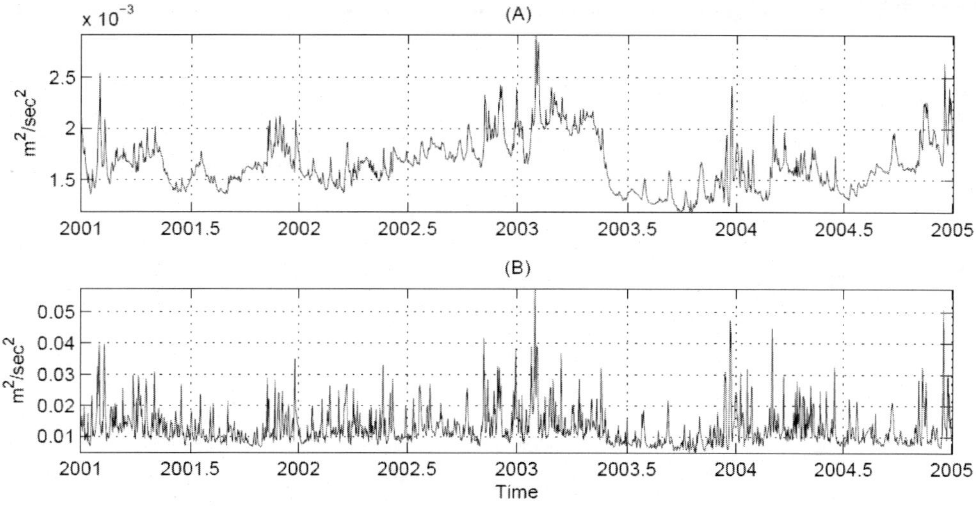

Figure 14. Daily time series of (A) total and (B) surface kinetic energy computed from SCRM simulations covering the period (2001-2004). Units are m2sec-2.

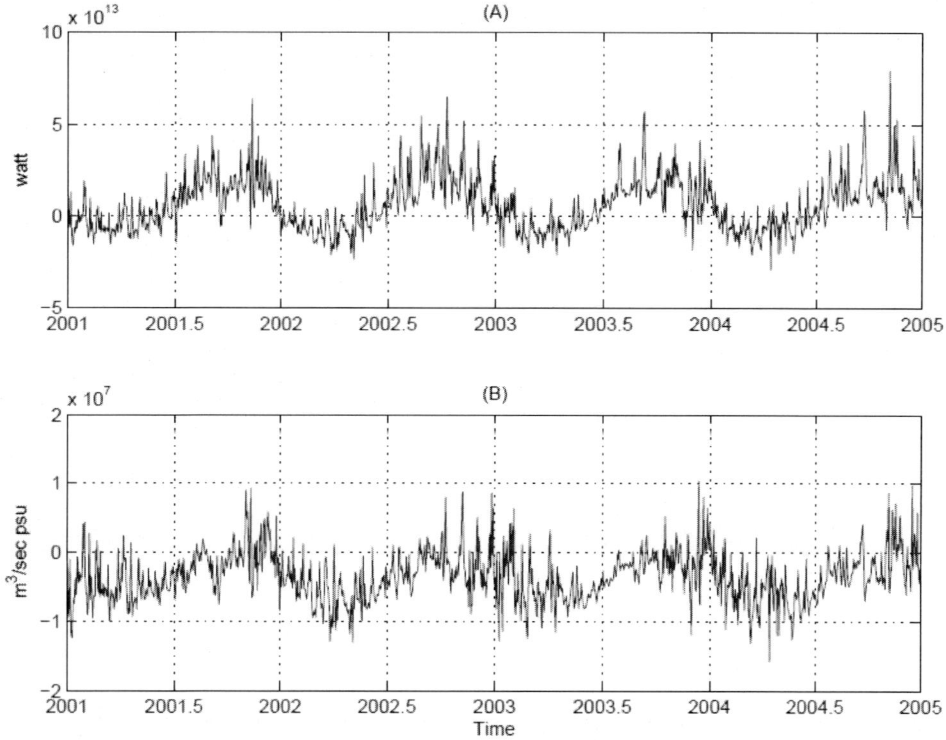

Figure 15. Time series of the net heat flux (A) and net salt flux at the Sicily Strait by the SCRM model during the simulated period (2001-2004).

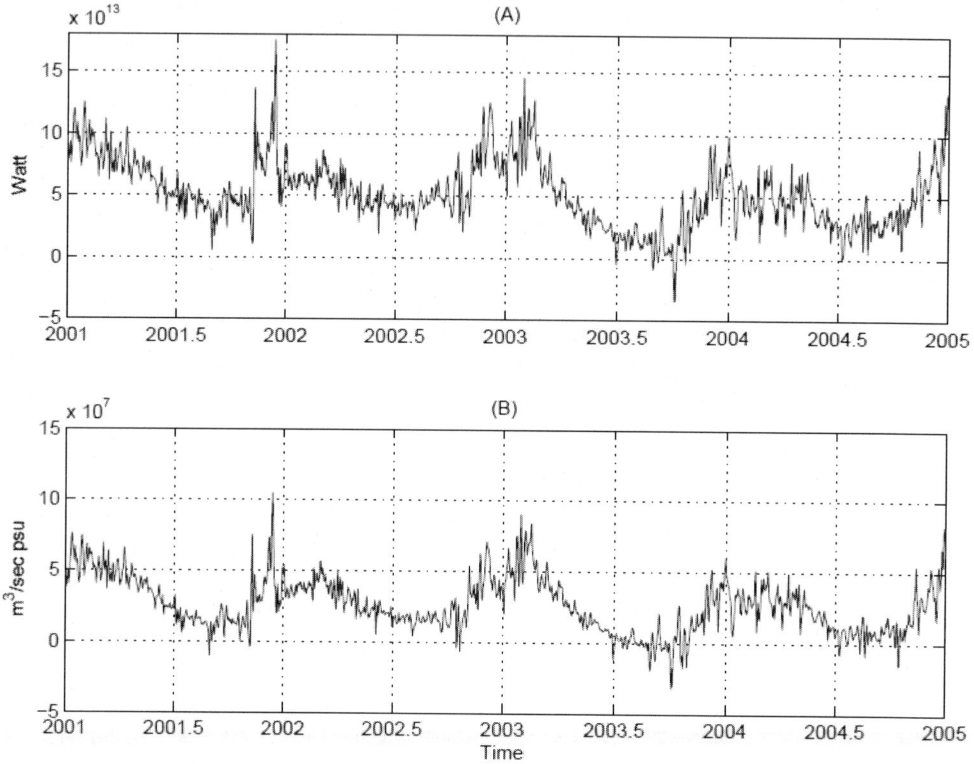

Figure 16. Time series of the net heat flux (A) and net salt flux at the Sardinia Channel by the SCRM model during the simulated period (2001-2004).

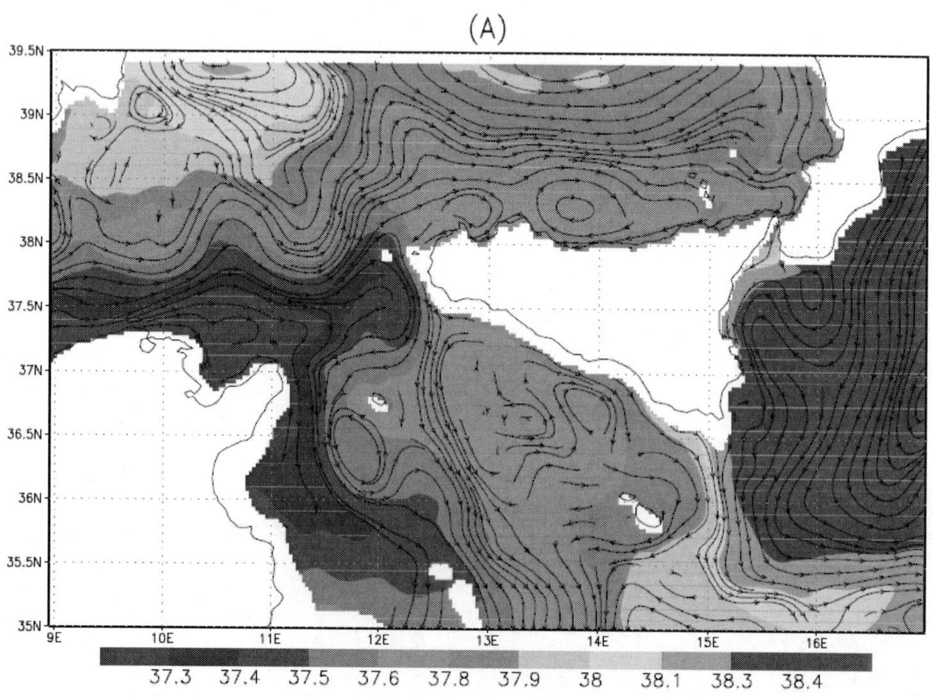

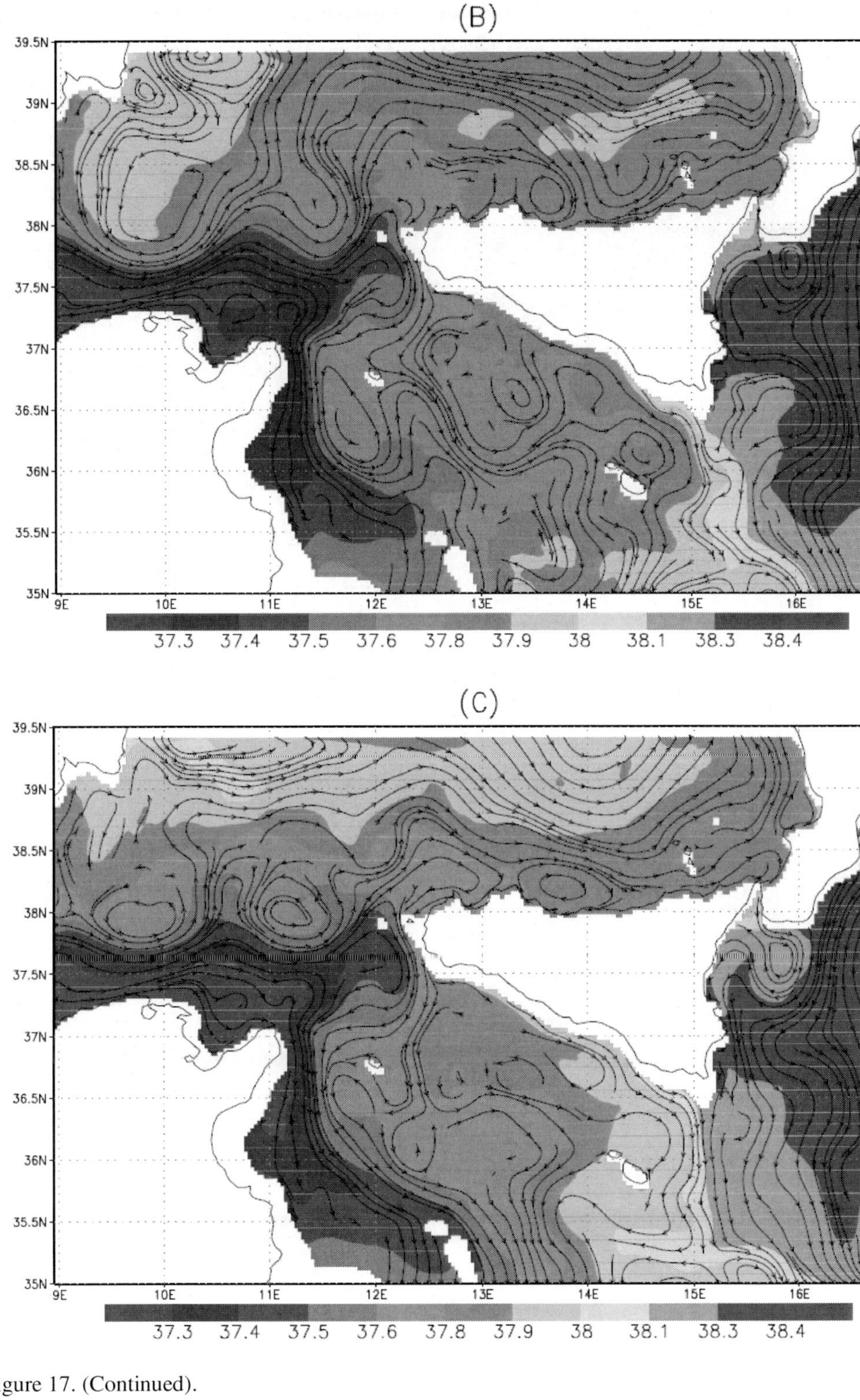

Figure 17. (Continued).

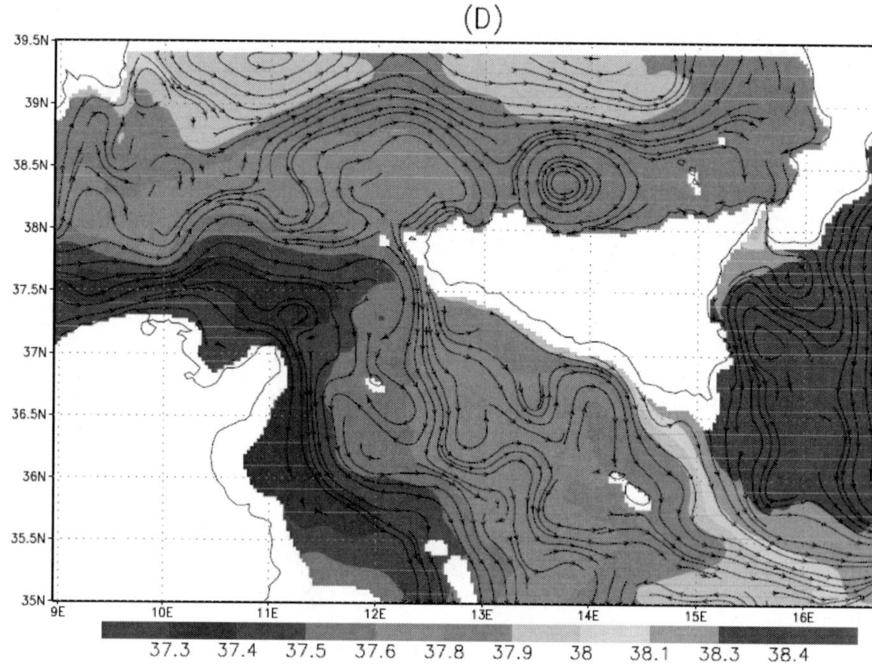

Figure 17. Near surface (5m) salinity (psu) superimposed to the velocity current for winter: (A) 2001; (B) 2002; (C) 2003; (D) 2004.

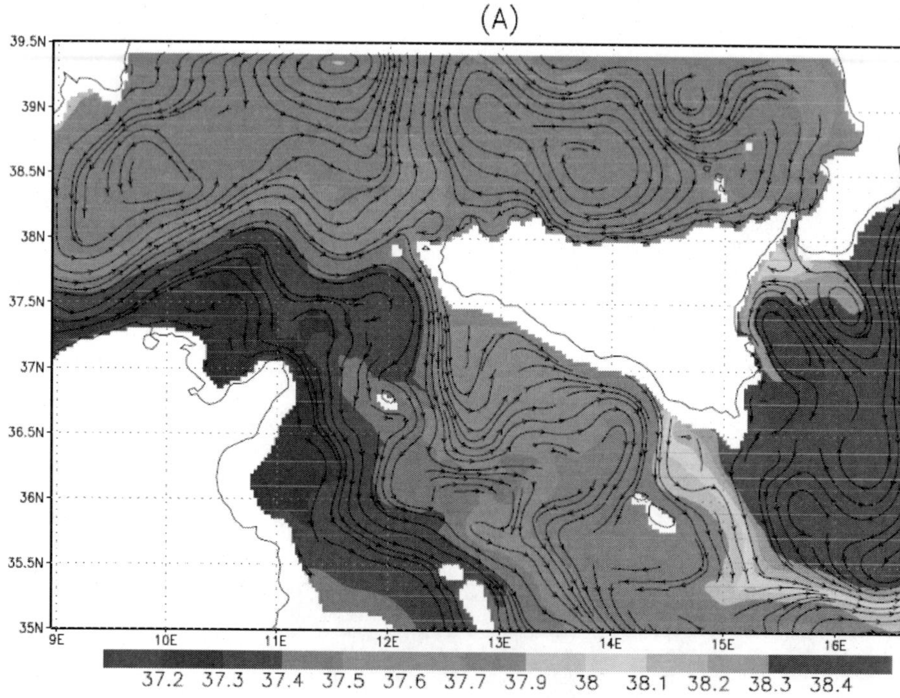

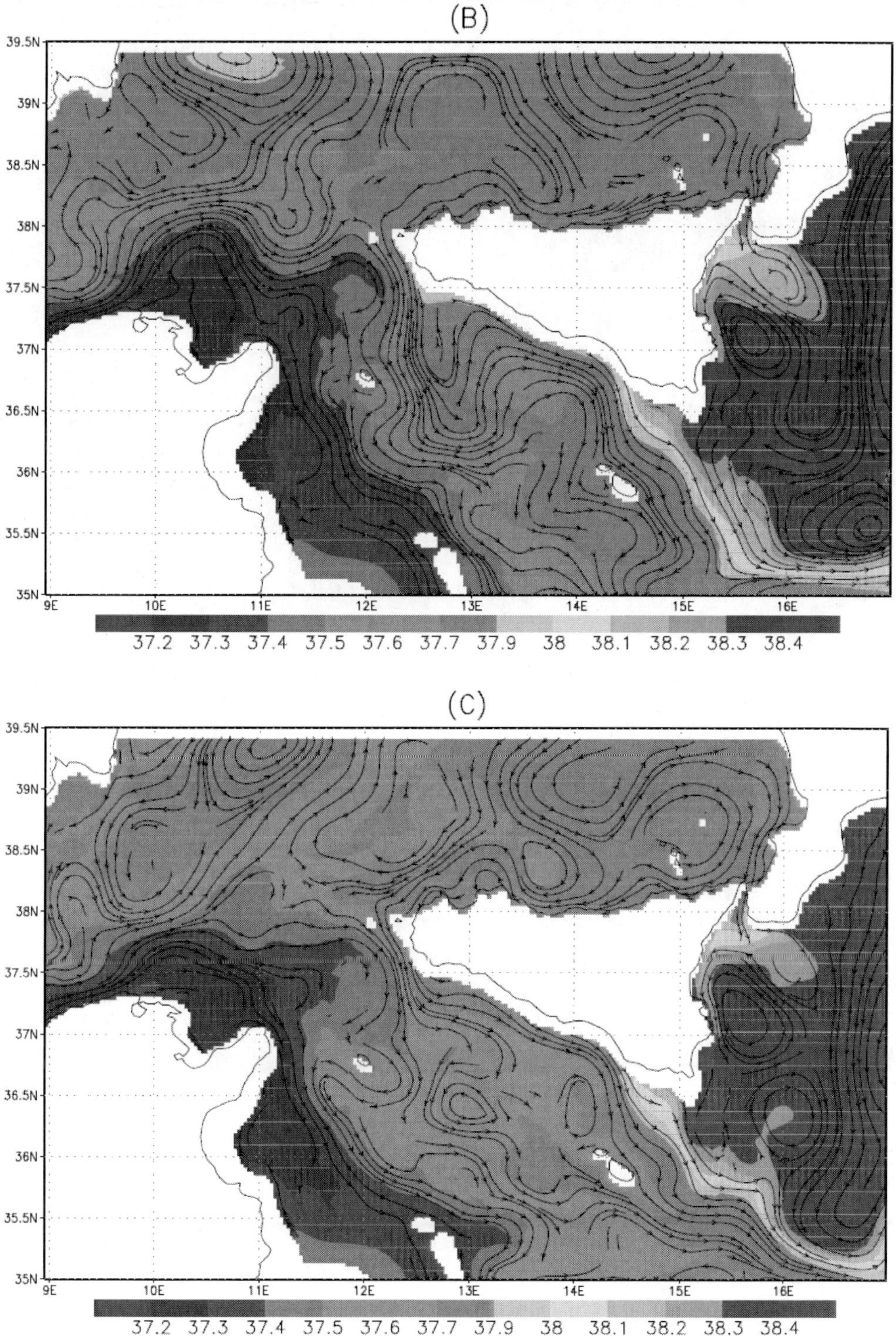

Figure 18. (Continued).

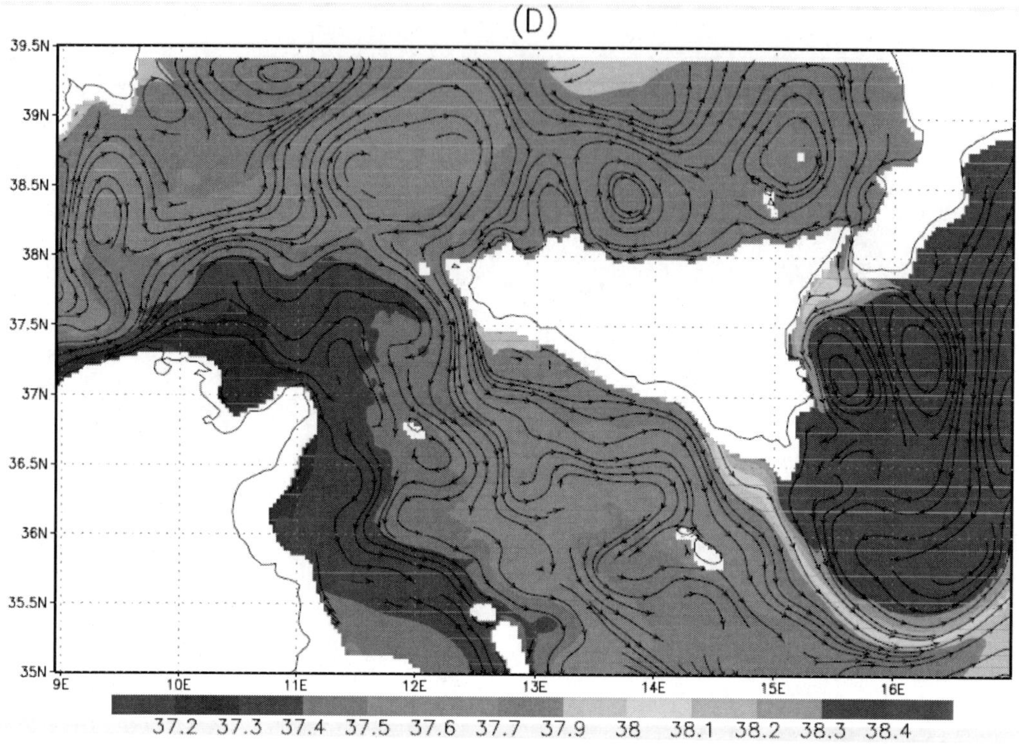

Figure 18. Near surface (5m) salinity (psu) superimposed to the velocity current for summer: (A) 2001; (B) 2002; (C) 2003; (D) 2004.

Concerning the salinity (Figures 17A-D and 18A-D respectively for winter and summer) the simulation reveals a sub-surface circulation dominated by the eastward advection of AW associated with jet stream currents like the AC, ATC, AIS and TBC. They are subject to significant spatial and temporal variability due to the influence of the wind stress, of the bottom topography and of the instability of the currents [20, 32]. This was evident through the generation of rather energetic mesoscale cyclonic and anti-cyclonic eddies, which can deviate the eastwards AW flow with important fluctuations in shape and strength of the current.

A deeper insight of the variability of surface hydrology and circulation can be accomplished decomposing their signal through EOF method, as in the next sub-section.

3.4. Modes of Variability

The model output can be used to assess the extent to which the seasonal cycle and observed anomalies bear on total variability. This is done by reducing the simulated dataset into a number of significant modes through the use of the EOF method. In this case the surface fields only are considered. Figures 19A-C show EOF1-3 of SST, respectively accounting for 98.8%, 0.89% and 0.15% of the total variance. In the first spatial mode (Figure 19A) the strongest signals are in the Tyrrhenian Sea, on the Ionian Trench and on the Tunisian shelf close to the Gulf of Gabes. On the other hand, the lowest values are along the southern Sicilian coast between the Adventure Bank and the Malta Channel due to the persistent presence of upwelling events that limit the range of thermal oscillation of the surface temperature. The first PC (Figure 20A) clearly indicates that the first mode explains most of the seasonal variability. As expected, being the SST highly auto-correlated and strongly dependent on the atmospheric forcing, the largest part (98%) of the total variance is related to the seasonal cycle. The map relative to the second spatial mode (Figure 20B), despite its low contribution to the total variance (0.89%), shows positive and negative structures located in the northern and the southern sectors with respect to the NW-SE axis of the Sicily Channel. This EOF mode integrates the signal given by a thermal anomaly centered and related to a heat wave experienced in Europe in June-August 2003. The weaker contribution of the third EOF mode (Figure 19C) is characterized by signals on the Tunisian shelf, while the behavior of the corresponding PC (Figure 20C) is associated with a strong interannual variability.

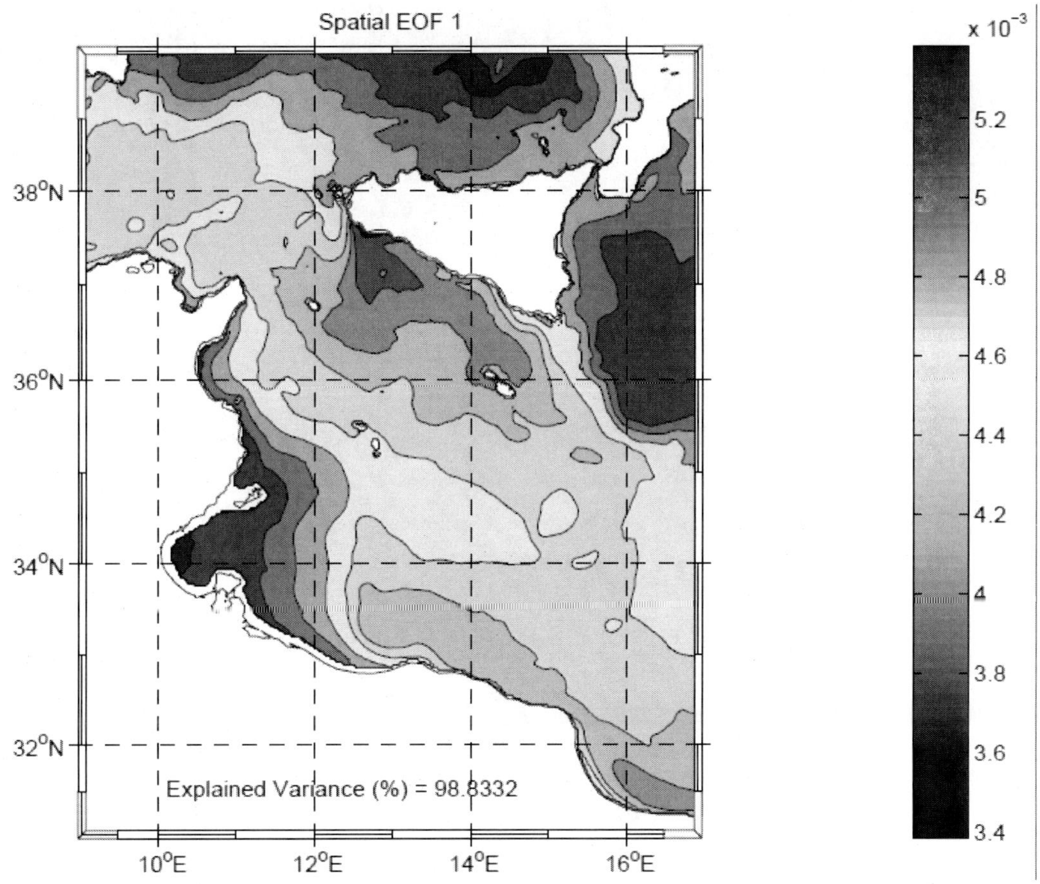

Figure 19. (Continued).

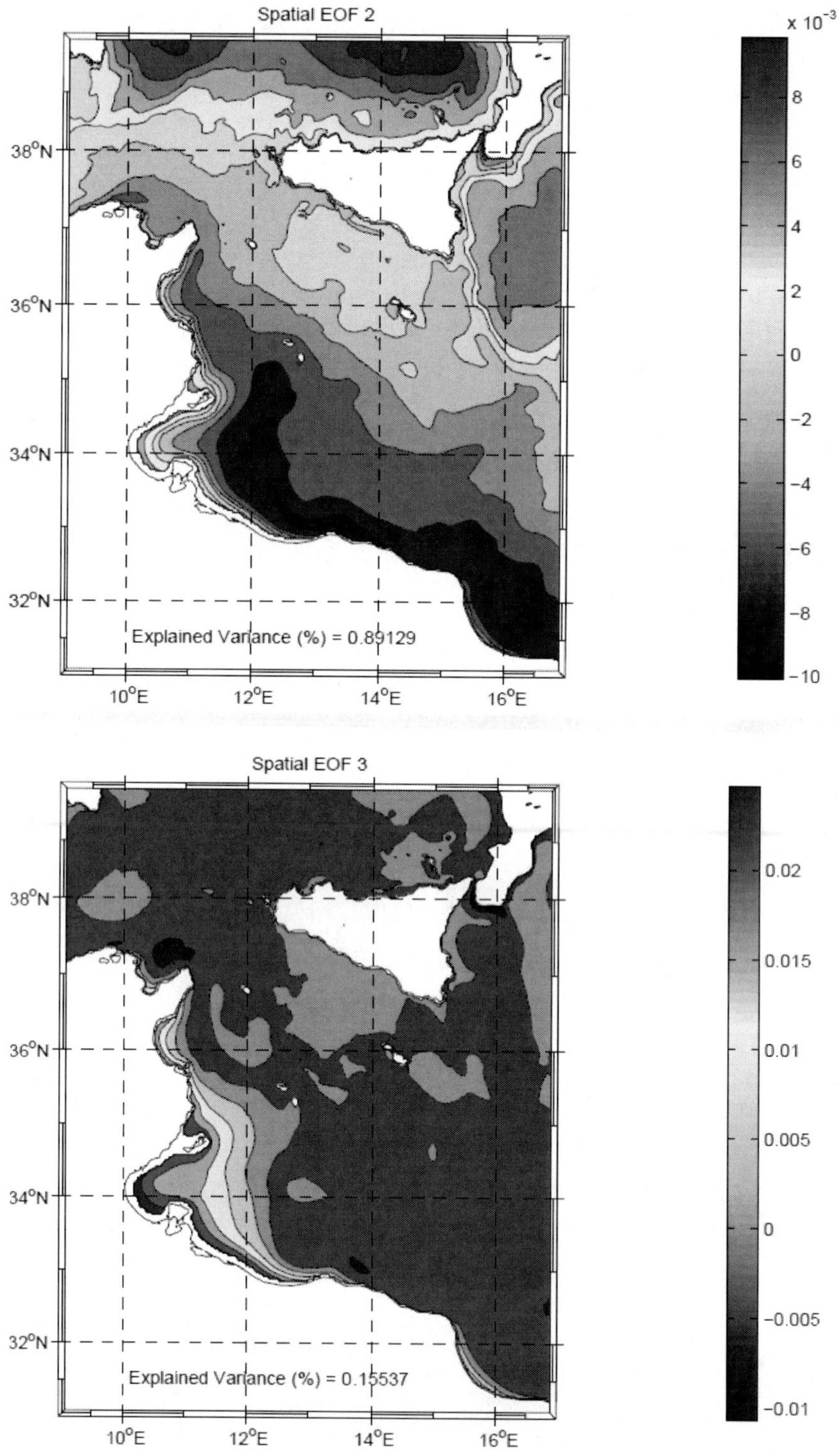

Figure 19. EOF-1(A), -2(B) and -3(C) of modeled sea surface temperature: the first mode explains more than 98% of the total variance of the dataset.

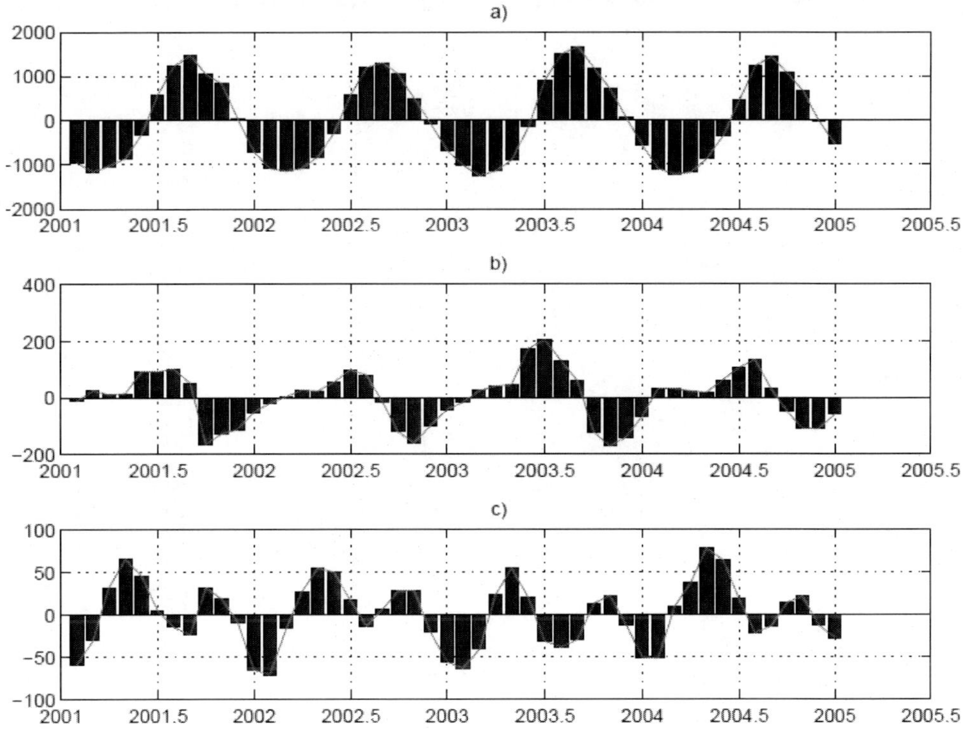

Figure 20. First three Principal Components of SST: top to bottom PC1 (A), PC2 (B) and PC3 (C).

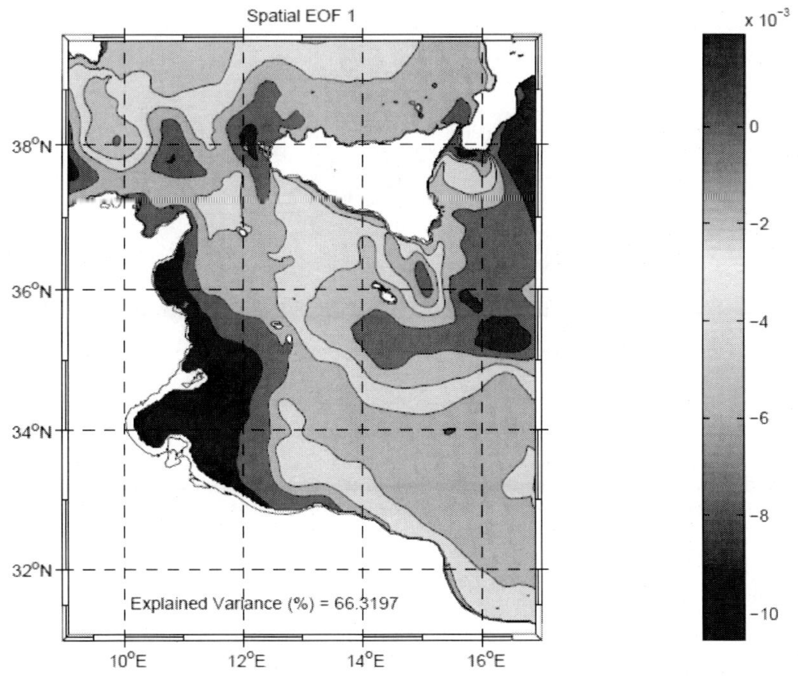

Figure 21. (Continued).

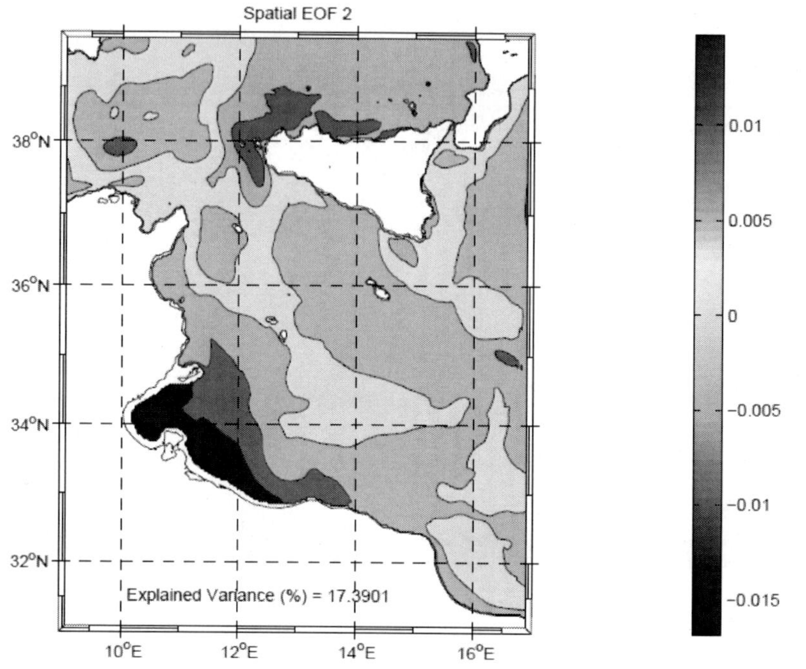

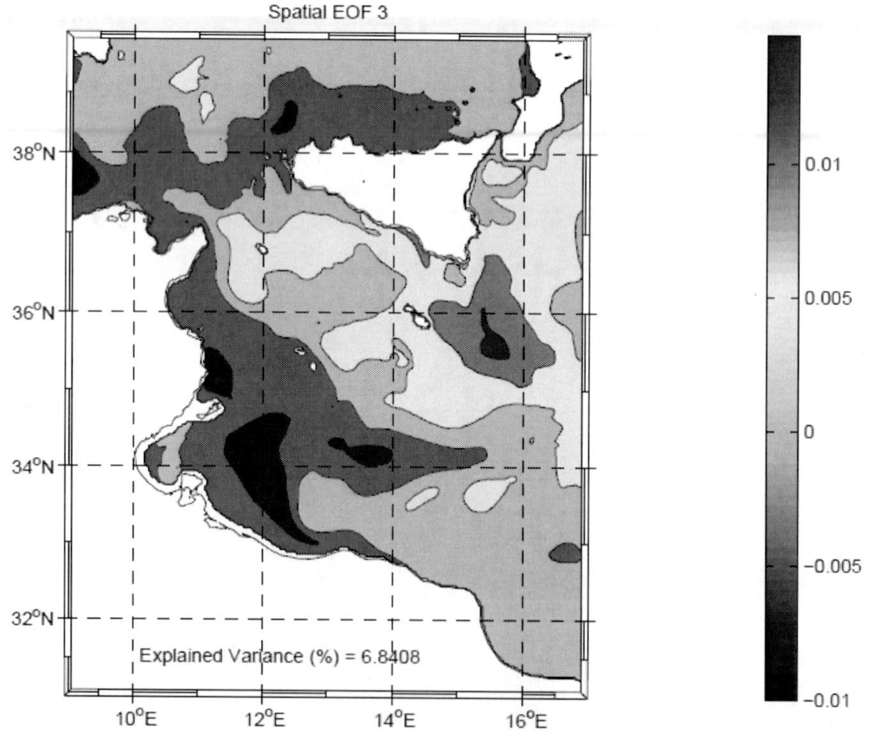

Figure 21. EOF-1(A), -2(B) and -3(C) of modelled sea surface salinity. First mode account for about 66% of total variance.

If compared with temperature, the salinity is a better tracer of the water mass displacements in the upper layers. EOF analysis has also been carried out for the surface salinity (Figures 21A-C) with the first three EOF modes explaining 66%, 17% and 7% of the total variance.

In the first spatial mode, the strongest signals (negative signs) are in the area of the Tunisian shelf, Sardinia Channel, and Tyrrhenian Sea in the form of disconnected areas along the parallel 38°N. This signal of variability is given by the pulse of the AW flux conveyed by the ATC along the Tunisian shelf and by the branch of AC moving to the Tyrrhenian Sea. On the other hand the Ionian area is characterized by positive spatial patterns, but with a lower variability for the presence of the Ionian type waters (warm and salty) moving south-westward along the eastern Sicilian coast to Cape Passero (extreme south-east of Sicily) [32], then contributing to the formation of a density front [72]. The temporal variability linked with spatial patterns described by the first EOF (66%) shows a well defined seasonal cycle (Figure 22A). This indicates that these patterns represent periodic structures, recurring annually and associated with the seasonal variation of the flux of Atlantic Water entering in the study area through the Sardinia Channel. The second spatial mode (Figure 21B; 17.39% of the variance) has its main signal in two areas showing a phase opposition, i.e., on the Tunisian shelf around the Gulf of Gabes and in the area north of Sicily with its highest values around the Adventure Bank (south-west of Sicily). These two structures have opposite signals with a maximum, in absolute value, centered in the Gulf of Gabes. These features could be attributed to the seasonal process of densification of the waters through the evaporation and the heat loss due to north-western cold and dry winds.

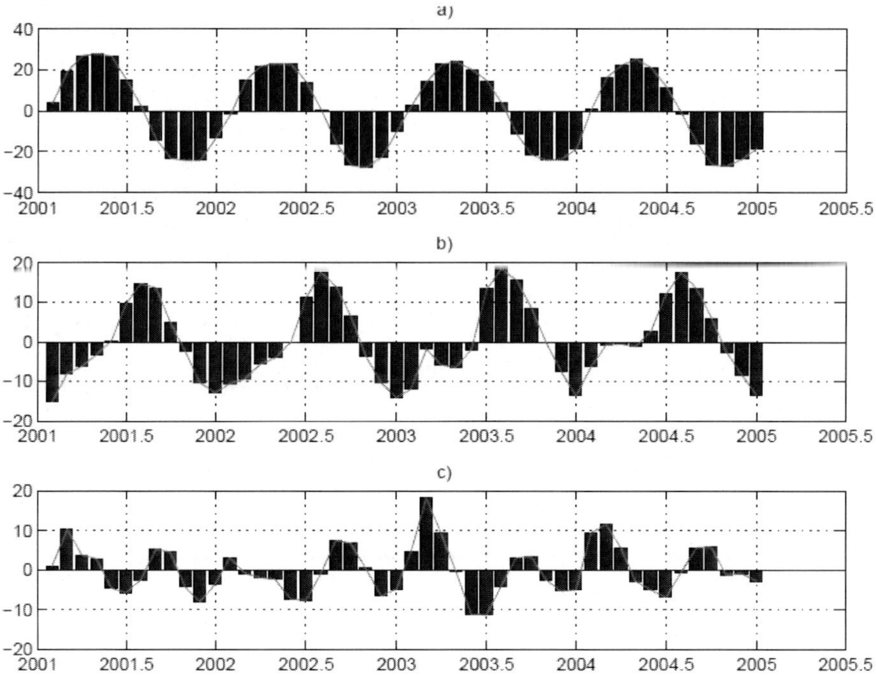

Figure 22. First three Principal Components of sea surface salinity: top to bottom PC1 (A), PC2 (B) and PC3 (C).

The third mode (Figure 21C), with 6.84% of the variance, shows features that are spatially in phase opposition. Negative values are associated to spatial patterns of the ATC and the TBC, while it's possible to see positive values in the Ionian basin, which describes the movement of the AIS. High values are associated with the Malta Channel Crest and on the Ionian slope, where the high variability of the AW flux linked to the transport of AIS is well known. Other patterns associated with the AIS, like the Adventure Bank Vortex and the Ionian Shelf Break Vortex, show a lower variability. In the figure 22C it's possible to observe the behavior of the relative PC; there is an anomaly, temporally localized between spring and summer 2003, which is associated with an anomaly in the AW transport through the Sardinia and the Sicily Channels, as drawn by the AW signature, clearly visible in the EOF-3. This signal is probably related to modifications of the AW flow across the Channels during the 2003 heat wave, as suggested by [72]. So, taking into account EOF-3 and respective PC, we may state that such modifications on AW flow occurred also during the spring preceding the heatwave. However, it could be necessary a more careful study to understand the possible correlation between AW flow anomalies and the enhancing of the heatwaves.

CONCLUSION

The space-time variability of the circulation in the Sicily and Sardinia Channels has been simulated from 1 January 2001 to 31 December 2004 through a three-dimensional, free surface, nested limited area ocean model having a horizontal resolution of 1/32°. An estimate of the variability associated to the seasonal cycle and interannual/anomalous signals has been carried out through the EOF decomposition of the simulated surface temperature and salinity. EOFs show that, as predicted, the annual cycle accounts for almost all the variance of surface temperature (~99%) mainly concentrated in the Tyrrhenian and Ionian seas, and in particular on the Tunisian shelf, while the variability is less emphasized in the Sicily and Sardinia Channels in correspondence with the upwelling areas. The thermal anomaly in summer 2003, included in the second mode of variability, accounts for less than 1% of total variance. Nevertheless this, it constitutes the main signal emerging from the series after removal of seasonality, thus confirming the importance and the amplitude of this extreme event.

The decomposition of variance in the surface salinity field has furnished a more complex picture of variability. The first two EOFs are both seasonal; the first EOF (66% of total variance) is mainly due to the periodicity of the AW inflow through the Sardinia Channel. This represents the dominant feature of the circulation characterized by rather energetic mesoscale cyclonic and anti-cyclonic eddies. The second EOF (17% of total variance) is related to the densification/evaporation processes occurring in coastal/shallow areas (mainly in the Gulf of Gabes). The main interannual signal of salinity, represented in the third mode (~7% of total variance), seems to be related to the anomalous behavior and path of the ATC, the BTC and the AIS, showing modifications of these well known circulation features.

The accuracy and quality of the model results has been estimated by using the RMS error as a diagnostic tool to compare numerical model fields with two independent observation sets: satellite SST and four CTD cruises. The comparison with the observed satellite data shows a minimum RMS error in winter (0.5°C) and a maximum in summer (1.5°C), during which the SCRM overestimates the observed satellite SST. Given that the SST satellite estimates have the lowest accuracy in the shallow coastal areas, the discrepancies might not be entirely due to the model imperfections. The comparison with the in-situ data shows an RMS error in the vertical profiles that does not exceed 1°C on average, apart from the observations in November 2002 and October 2004 where the two profiles are characterized by a subsurface maximum error that reaches about 2.5°C, while at the surface and deeper levels the RMS error remains below 0.8°C. These errors could be due to several causes: the computation of the diffusion terms, the inaccurate physics of the bulk formulae, the missing vertical numerical resolution attributed to the thickening of the first sigma layer in deep regions, and an inaccurate vertical mixing affecting the upper layers, especially when the water column is stratified in summer and fall. Salinity RMSE profiles are also influenced by the water column stratification, and particularly by the subsurface vertical gradient between the AW and the LIW. The salinity error profile shows an increasing trend from the surface (~ 0.25) at the transitional layer where the maximum values range from 0.4 to 0.8 in January 2004 when the transitional layer is strong. The origin of this error can be due to the SCRM smoothing the vertical profile at the transitional layer between AW and LIW, as well as through the dislocation of AW and LIW cores arising in consequence of an excessive vertical mixing with the surrounding water masses.

In conclusion, the skill of the SCRM model has been proved to match synoptic SST fields observed by satellites, as well as to reproduce the water column structure when and where such a data are available. By means of modeling data and statistical analyses, like the EOF decomposition, it has been possible to identify and to quantify the spatial and temporal variability of the main sub-basin and mesoscale features occurring in the study area and their anomalous behavior. The obtained results have also furnished valid suggestions in order to address further research and modeling efforts. Major mismatches of the model are found at the interface between different water masses and when/where the surface mixed layer becomes a dominant component of the surface dynamics. This evidence has suggested that further efforts have to be accomplished to better describe the vertical mixing, especially at the surface, and better resolve better the different water masses in the vertical dimension.

ACKNOWLEDGMENTS

The authors thank Mr Mireno Borghini from ISMAR-CNR in La Spezia for his contribution on instrument and data quality checking, the captains and the crews onboard the R/V Urania during the four MedGOOS cruises. This research was partially supported by the Italian national research project MUR-PON SIGLA and the EU SPF INDOO (EU Contract Number ASIE/2005/102-483). The hydrodynamic model SCRM has been developed in the framework of the EU MFSTEP – Mediterranean Forecasting System Toward Environmental Prediction (EU Contract Number EVK3-CT-2002-00075).

REFERENCES

[1] Astraldi, M., Conversano, F., Civitarese G., Gasparini G. P., Ribera d'Alcalà M. & Vetrano A. (2002a). Water mass properties and chemical signatures in the central Mediterranean region. *J. Marine Systems*, 33-34, 155-177.

[2] Millot, C. (1999). Circulation in the Western Mediterranean Sea. *J. Marine Systems, 20*, 423-442.

[3] Pinardi, N., Korres, G., Lascaratos, A., Roussenov, V. & Stanev, E. (1997). Numerical simulation of the interannual variability of the Mediterranean Sea upper ocean circulation. *Geophysical Research Letters*, *24*, 4, 425-428.

[4] Robinson, A. R., Golnaraghi, M., Leslie, W. G., Artegiani, A., Hecht, A., Lazzoni, E., Michelato, A., Sansone, E., Theocharis, A. & Umluata, U. (1991). The eastern Mediterranean general circulation: features, structure and variability. *Dynamics of Atmospheres and Oceans*, *15*, 3-5, 215-240.

[5] Piccioni, A., Gabriele, M., Salusti, E. & Zambianchi, E. (1988). Wind-induced upwellings off the southern coast of Sicily. *Oceanologica Acta*, *11*, 4, 309-314.

[6] Beranger, K., Mortier, L., Gasparini, G. P., Gervasio, L., Astraldi, M. & Crepon, M. (2004). The dynamics of the Sicily Strait: a comprehensive study from observations and models. *Deep-Sea Research II*, *51*, 411-440.

[7] Robinson, A. R., Sellschopp, J., Warn-Varnas, A., Leslie, W. G., Lozano, C. J., Haley Jr. P. J., Anderson L. A. & Lermusiaux, P. F. J. (1999). *The Atlantic Ionian Stream. J. Marine Systems*, *20*, 129-156.

[8] Grancini, G. F. & Michelato, A. (1987) Current structure and variability in the Strait of Sicily and adjacent area. *Annales Geophysicae*, 5B, 1, 75-88.

[9] Manzella, G. M. R., Hopkins, T. S., Minnett, P. J. & Nacini, E. (1990). Atlantic Water in the Strait of Sicily. *J. Geophysical Research*, *95*, C2, 1569-1575.

[10] Sammari, C., Millot, C., Taupier-Letage, I., Stefani, A. & Brahim, M. (1999). Hydrological characteristics in the Tunisia-Sardinia-Sicily area during spring. *Deep-Sea Research,* I, *46*, 1671-1703.

[11] Astraldi, M., Balopoulos, S., Candela, J., Font, J., Gacic, M., Gasparini, G. P., Manca, B., Theocharis, A. & Tintore, J. (1999). The role of straits and channels in un derstanding the characteristics of Mediterranean circulation. *Progress in Oceanography*, *44*, 65-108.

[12] Astraldi, M., Gasparini, G. P., Vetrano, A. & Vignudelli, S. (2002b). Hydrographic characteristics and interannual variability of water masses in the central Mediterranean: a sensitivity test for long-term changes in the Mediterranean Sea. *Deep-Sea Research*, I, *49*, 661-680.

[13] Vetrano, A., Gasparini, G. P. & Astraldi, M. (2004). Surface circulation and fluxes inside the Central Mediterranean Sea In: CIESM (Ed.), *Rapports de la Commission Internationale pour l'Exploration Scientifique de la mer Mediterranee*, *37*, 149.

[14] Gasparini, G. P., Ortona, A., Budillon, G., Astraldi, M. & Sansone, E. (2005). The effect of the Eastern Mediterranean Transient on the hydrographic characteristics in the Strait of Sicily and in the Tyrrhenian Sea. *Deep-Sea Research*, I, *52*, 915-935.

[15] Manzella, G. M. R., Gasparini, G. P. & Astraldi, M. (1988). Water exchange between the eastern and western Mediterranean through the Strait of Sicily. *Deep-Sea Research,* I, *35*, 1021–1035.

[16] Zavatarelli, M. & Mellor, G. L. (1995). A numerical study of the Mediterranean Sea circulation. *J. Physical Oceanography*, *25*, 1384-1414.

[17] Lascaratos, A., Roether, W., Nittis, K. & Klein, B. (1999) Recent changes in deep water formation and spreading in the eastern Mediterranean Sea: a review. *Progress in Oceanography*, *44*, 1-3, 5-36.

[18] Beranger, K., Mortier, L. & Crepon, M. (2005). Seasonal variability of water transport through the Straits of Gibraltar, Sicily and Corsica, derived from a high-resolution model of the Mediterranean circulation. *Progress in Oceanography*, *66*, 341-364.

[19] Pierini, S. & Rubino, A. (2001). Modeling the Oceanic Circulation in the Area of the Strait of Sicily: The Remotely Forced Dynamics. *J. Physical Oceanography*, *31*, 1397-1412.

[20] Lermusiaux, P. F. J. & Robinson, A. R. (2001) Features of dominant mesoscale variability, circulation patterns and dynamics in the Strait of Sicily. *Deep-Sea Research,* I, *48*, 1953-1997.

[21] Molcard, A., Gervasio, L., Griffa, A., Gasparini, G. P., Mortier, L. & Tamay, M. O. (2002). Numerical investigation of the Sicily Channel dynamics: density currents and water mass advection. *J. Marine Systems*, *36*, 219-238.

[22] Pinardi, N., Allen, I., Demirov, E., De Mey, P., Korres, G., Lascaratos A., Le Traon, P. Y., Maillard, C., Manzella, G. M. R. & Tziavos, C. (2003). The Mediterranean ocean forecasting system: first phase of implementation. *Annales Geophysicae*, *21*, 3-20.

[23] Sorgente, R., Drago, A. F. & Ribotti, A. (2003). Seasonal variability in the Central Mediterranean Sea circulation. *Annales Geophysicae*, *21*, 299-322.

[24] Oddo, P., Pinardi N. & Zavatarelli, M. (2005). A Numerical Study of the Interannual Variability of the Adriatic Sea (2000-2002). *Science of the Total Environment*, *353*, 1-3, 39-56.

[25] Drago, A. F., Sorgente, R. & Ribotti, A. (2003). A high resolution hydrodynamic 3-D model simulation of the Malta shelf area. *Annales Geophysicae*, *21*, Special Issue, Mediterranean Forecasting System Pilot Project (MFSPP), 323 344.

[26] Puillat, I., Sorgente, R. & Ribotti, A. (2003). Circulation in the Sardinian Sea: a synergetic approach by in situ data, a three dimensional model and infrared imagery. *Geophysical Research Abstracts*, *5*, 13117.

[27] Onken, R., Robinson, A. R., Lermusiaux, P. F. J., Haley, Jr. P. J. & Anderson, L. A. (2003). Data-driven simulations of synoptic circulation and transports in the Tunisia-Sardinia-Sicily region. *J. Geophys. Research, 108*, C9, 1-13.

[28] Ribotti, A., Puillat, I., Sorgente, R. & Natale, S. (2004). Mesoscale circulatio918 n in the surface layer off the southern and western Sardinia island in 2000-2002. *Chemistry and Ecology*, *20*, 5, 345-363.

[29] Hopkins, T. S. (1988). Recent observations on the intermediate and deep water circulation in the Southern Tyrrhenian Sea. In: Minas HJ, Nival P (Ed.), Oceanographie pelagique mediterraneenne, *Oceanologica Acta*, SP, 41-50.

[30] Krivosheya, V. G. (1983). Winter circulation and structure in the Tyrrhenian Sea. *Oceanology*, *23*, 2, 166-171.

[31] Herbaut, C., Codron, F. & Crepon, M. (1998). Separation of a Coastal Current at a Strait Level: Case of the Strait of Sicily. *Journal of Physical Oceanography, 28*, 1346-1362.
[32] Beranger, K., Mortier, L., Gasparini, G. P., Gervasio, L., Astraldi, M. & Crepon, M. The dynamics of the Sicily Strait: a comprehensive study from observations and model, *Deep Sea Res. II, 51*, 411–440, 2004.
[33] Astraldi, M., Gasparini, G. P., Sparnocchia, S., Moretti, M. & Sansone, R. (1996). The characteristics of the water masses and the water transport in the Sicily Channel at long time scales. In: Briand F. (Ed.), *Dynamics of Mediterranean Straits and Channels. Bulletin de l'Institut Oceanographique, CIESM Science Series, 2*, 17, 95-116.
[34] Malanotte-Rizzoli, P., Manca, B. B., Ribera d'Alcalà, M., Theocharis, A., Bergamasco, A., Bregant, D., Budillon, G., Civitarese, G., Georgopoulos, D., Michelato, A., Sansone, E., Scarazzato, P. & Souvermezoglou, E. (1997). A synthesis of the Ionian Sea hydrography, circulation and water masses pathways during POEM-Phase I. *Progress in Oceanography, 39*, 153-204.
[35] Gasparini, G. P., Ortona, A., Budillon, G., Astraldi, M. & Sansone, E. (2004). The propagation of the eastern Mediterranean transient from the eastern to the western Mediterranean basin In: CIESM (Ed.), *Rapports de la Commission Internationale pour l'Exploration Scientifique de la mer Mediterranee, 37*, 101.
[36] Poulain, P. M. & Zambianchi, E. (2007). Surface circulation in the central Mediterranean Sea as deduced from Lagrangian drifters in the 1990s, *Continental Shelf Research, 27*, 981–1001.
[37] Hamad, N., Millot, C. & Taupier-Letage, I. (2005). A new hypothesis about the surface circulation in the eastern basin of the Mediterranean sea. *Progress in Oceanography, 66*, 2-4, 287-298.
[38] Marullo, S., Santoleri, R., Malanotte-Rizzoli, P. & Bergamasco, A. (1999). The sea surface temperature field in the eastern Mediterranean from AVHRR data. Part I: Seasonal variability. *J. Marine Systems, 20*, 1-4, 63-81.
[39] Pedlosky, J. (1992). Geophysical Fluid Dynamics. Springer New York, 710 pp.
[40] Dhatt, G., Lefrançois, E. & Touzot, G. (2012). Finite Element Method. John Wiley & sons, 612 pp.
[41] Sorgente, R., Olita, A., Oddo, P., Fazioli, L. & Ribotti, A. (2011). Numerical simulation and decomposition of kinetic energy in the central Mediterranean: insight on mesoscale circulation and energy conversion, *Ocean Science*, vol *7*, 503–519.
[42] Blumberg, A. F. & Mellor, G. (1987). A description of a three-dimensional coastal ocean circulation model. In: Heaps NS (Ed.). Three-dimensional Coastal Ocean Models. *Coastal Estuarine Science,* AGU, 1-16.
[43] Mellor, G. L. (1991). An equation of state for numerical models of oceans and estuaries. *J. Atmos. Oceanic Tech., 8*, 609-611.
[44] Mellor, G. L. & Yamada, T. (1982). Development of a turbulence closure model for geophysical fluid problems. *Rev. Geophys. Space Phys., 20*, 851-875.
[45] Smagorinsky, J. (1993). Some historical remarks on the use of nonlinear viscosities. In: Galperin B., Orszag S.A. (Eds.), Lare eddy simulation of complex engineering and geophysical flows, *Cambridge Univ. Press*, 3-36 pp.
[46] Mellor, G. L., Ezer, T. & Oey, L. Y. (1994). The pressure gradient conundrum of sigma coordinate ocean models. *J. Atmos. Oceanic. Tech., 11*, 1126-1134.

[47] Auclair, F., Marsaleix, P. & Estournel, C. (2000). Sigma coordinate pressure gradient errors: evaluation and reduction by an inverse method. J. *Atmospheric and Oceanic Technology*, *17*, 1348-1367.

[48] Gabersek, S., Sorgente, R., Natale, S., Ribotti, A., Olita, A., Astraldi, M. & Borghini, M. (2007). The Sicily Channel Regional Model forecasting system: initial boundary conditions sensitivity and case study evaluation. *Ocean Science Special Issue*, *3*, 31-41.

[49] Auclair, F., Marsaleix, P. & Estournel, C. (2001). The penetration of the Northern Current over the Gulf of Lions (Mediterranean) as a downscaling problem. *Oceanologica Acta*, *24*, 6, 529-544.

[50] Korres, G. & Lascaratos, A. (2003). A one-way nested eddy resolving model of the Aegean and Levantine basins: implementation and climatological runs. *Annales Geophysicae*, *21*, *Special Issue on Mediterranean Forecasting System Pilot Project (MFSPP),* 205-220.

[51] Zavatarelli, M. & Pinardi, N. (2003). The Adriatic Sea modelling system: a nested approach. Annales Geophysicae, 21, *Special Issue on Mediterranean Forecasting System Pilot Project* (MFSPP), 345-364.

[52] Hellerman, S. & Rosenstein, M. (1983). Normal monthly wind stress over the world ocean with error estimates. *J. Phys. Oceanogr.*, *13*, 1093-1104.

[53] Reed, R. K. (1975). An evaluation of formulas for estimating clear-sky insolation over the ocean. *NOAA Tech. Rep. ERL 352-PMEL*, *26*, 25 pp.

[54] Reed, R. K. (1977). On estimating insolation over the ocean. *J. Phys. Oceanogr. 7*, 482-485.

[55] Bignami, F., Marullo, S., Santoleri, R. & Schiano, M. E. (1995). Longwave radiation budget in the Mediterranean Sea. *J. Geophysical Research-Oceans*, *100*, 2501-2514.

[56] Kondo, J. (1975). Air-sea bulk transfer coefficients in diabatic conditions. *Bound. Layer Meteor.*, *9*, 91-112.

[57] Legates, D. R. & Wilmott, C. J. (1990). Mean seasonal and spatial variability in gauge corrected global precipitation. *Int. J. Climatol.*, *10*, 111-127.

[58] Schlitzer, R. (2007). Ocean Data View, http:odv.awi.de.

[59] Lorenz, E. (1956). Empirical orthogonal functions and statistical weather prediction. Tech. Rep. 1, Statistical Forecasting Project, Department of Meteorology, Massachusetts Institute of Technology, Cambridge, MA, 49 pp.

[60] Bartolacci, D. M. & Luther, M. E. (1999). Patterns of co-variability between physical and biological parameters in the Arabian sea. *Deep-Sea Research II*, *46*, 1933–1964.

[61] Baldacci, A., Corsini, G., Grasso, R. G. M., Allen, J. T., Cipollini, P., Guymer, T. H. & Snaith, H. M. (2001). A study of the Alboran sea mesoscale system by means of empirical orthogonal function decomposition of satellite data. *Journal of Marine Systems*, *29*, 293–311.

[62] Bouzinac, C., Font, J. & Johnny, J. (2003). Annual cycles of sea level and sea surface temperature in the western Mediterranean sea. *Journal of Geophysical Research*, *108*, C03, 1-20.

[63] Olita, A., Ribotti, A., Sorgente, R., Fazioli, L. & Perilli, A. (2011). SLA-chlorophyll-a variability and covariability in the Alghero-Provençal Basin (1997-2007) through combined use of EOF and wavelet analysis of satellite data. *Ocean Dynamics*, *61*, 89-102. DOI 10.1007/s102036-010-0344-9.

[64] Emery, W. J. & Thompson, R. E. (1997). Data analysis in physical oceanography. Pergamon, New York.

[65] Simmons, A. J., Branstator, G. W. & Wallace, J. M. (1983). Barotropic wave propagation, instability and atmospheric teleconnection patterns. *J. Atmos. Sci.*, *40*, 1363-1392.

[66] Von Storch, H. & Zwiers, F. W. (2001). Statistical analysis in climate research. *Cambridge University Press*, 484 pp.

[67] Preisendorfer, R. W. (1988) Principal Component Analysis in meteorology and oceanography. Elsevier, Amsterdam, 425 pp.

[68] Chai, T. & Draxler, R. R. (2014). Root mean square error (RMSE) or mean absolute error (MAE)? – Arguments against avoiding RMSE in the literature. *Geosci. Model Dev.*, *7*, 1247-1250, 2014. doi:10.5194/gmd-7-1247-2014.

[69] Chiggiato, J. & Oddo, P. (2008) Operational ocean models in the Adriatic Sea: a skill assessment. *Ocean Science*, *4*, 1, 61-71.

[70] Onken, R., lvarez, A., Fernndez, V., Vizoso, G., Basterretxea, G., Tintor, J., Haley, P. & Nacini, E. (2008). A forecast experiment in the Balearic Sea. *Journal of Marine Systems*, *71*, 1-2, 79-98.

[71] Black, E., Blackburn, M., Harrison, G., Hoskins, B. & Methven, J. (2004). Factors contributing to the summer 2003 European Heatwave. *Weather*, *59*, 8, 217-222.

[72] Olita, A., Sorgente, R., Natale, S., Gabersek, S., Ribotti, A., Bonanno, A. & Patti, B. (2007). Effects of the 2003 European heatwave on the Central Mediterranean Sea: surface fluxes and the dynamical response. *Ocean Science*, *3*, 273-289.

In: Geophysics
Editor: Gemma Aiello

ISBN: 978-1-63484-831-2

Chapter 6

MONITORING TEST OF CRACK OPENING IN VOLCANIC TUFF (COROGLIO CLIFF, ITALY) USING DISTRIBUTED OPTICAL FIBER SENSOR

Aldo Minardo[1], Mauro Caccavale[2], Agnese Coscetta[1], Giuseppe Esposito[2], Fabio Matano[2,*], Marco Sacchi[2], Renato Somma[3], Giovanni Zeni[4] and Luigi Zeni[1]

[1]Department of Industrial and Information Eng., Second University of Naples, Aversa, Italy
[2]Institute for Marine and Coastal Environment (IAMC), Italian National Research Council (CNR), Naples, Italy
[3]National Institute of Geophysics and Volcanology (INGV), Naples Section – Osservatorio Vesuviano, Naples, Italy
[4]Institute for Electromagnetic Sensing of the Environment (IREA), Italian National Research Council (CNR), Naples, Italy

ABSTRACT

We report the experimental application of distributed optical fiber sensors based on stimulated Brillouin scattering (SBS) through the so-called Brillouin Optical Time Domain Analysis (BOTDA) to the monitoring of artificially induced crack opening in a volcanic rock slope. The aim of this chapter is to show how the sensing optical fiber cable is able to detect the formation and to follow the growth of fractures in the tuff rocks and to identify their location along the cliff. The experiments have been performed at the base of the Coroglio tuff cliff in Naples (Italy) and have demonstrated that early detection of crack opening can be obtained and development of early warning systems is an attainable goal of the research.

Keywords: fiber optic sensors, BOTDA, landslide monitoring, rock slope failures, tuff cliff

* Corresponding Author address: CNR-IAMC, Calata Porta di Massa, Porto di Napoli, 80133, Napoli, ITALY. Email: fabio.matano@cnr.it.

INTRODUCTION

During recent years the distributed optical fiber sensors have gained a considerable attention in structural and environmental monitoring due to their specific advantages. The classical advantages, such as immunity to electromagnetic interferences, high sensitivity, small size and possibility to be embedded into the structures, multiplexing and remote interrogation capabilities [1], are common to all optical fiber sensors. The peculiar advantages of distributed optical fiber are represented by the unique feature of allowing the exploitation of a common (i.e., telecommunication grade) optical fiber cable as the sensing element to measure deformation and temperature profiles over very long distances.

In the described approach, distributed optical fiber sensors based on Stimulated Brillouin Scattering (SBS), through the so-called Brillouin Optical Time Domain Analysis (BOTDA), have been employed [2]. They have already been successfully exploited in the monitoring of large civil and geotechnical structures such as bridges, tunnels, dams, pipelines and so on, allowing to identify and to localize any kind of failures that can occur during their construction and operation [3, 4].

The landslide monitoring is a target for current research in the fields of geology, hydrogeology and geodesy. Landslides are regarded as a major geologic hazard and cause every year severe damage to urbanized areas and/or infrastructures. The availability of remote monitoring systems characterized by real-time capabilities and flexibility is mandatory in order to reduce the impact from hazardous landslides. The continuous data provided by real-time monitoring allow for an immediate detection of landslide activity, that can be crucial in making timely decisions about safety and provide a better understanding of the dynamic landslide behavior that, in turn, enables engineers to create more effective designs to prevent or to minimize the effects of landslides.

The causes, speeds and potential destructiveness of landslides widely vary, so there is no standard monitoring set-up that will universally work [5]. Some landslides move slowly, traveling only a few centimeters in many days or months. Other landslides can move suddenly, as mud or debris flows, that travel hundreds to thousands of meters in a few minutes and cause massive destruction and fatalities. Many landslides move only during or after long-lasting periods of infiltration of the soils by rain or melting snow, when groundwater pressures increase [6].

The sensors may be installed to detect precipitation, soil moisture and groundwater pressures, and/or slide displacement and acceleration, strongly depending on field conditions. The measurement over time of superficial ground displacements represents one of the most effective means for slope failure prediction among all the measurable parameters linked to the slope movements [7, 8]. The conventional geodetic and geotechnical instruments, such as the total stations, the GPS receivers and the extensometers, do not properly match the emergency requirements during slope failures. Indeed, they are able to provide precise measurements only over a few points and not over the wide sectors of the unstable area.

The distributed Brillouin-based sensing systems are capable of measuring the strain along a dedicated optical fiber cable fixed to the structure or to the slope to be monitored. The readings of the measurements can be taken every ten centimeters. This is a clear advantage when localized fractures need to be detected, but their position is not known beforehand. In this paper we present the application of BOTDA to the monitoring of movements along cracks in a rocky slope, made of volcanic tuffs.

The coastal cliffs are characterized by a rapid geomorphological evolution due to intense weathering, erosion and failure processes, which cause abrupt changes in the shape of the slope. The geomorphic evolution of cliffs is characterized by a set of failures including rock falls, rock topples and rock slides. In this context, the rock discontinuities play a fundamental role, as they exert a significant control on the shape, the volume and the failure susceptibility of detached rock blocks.

The coastal cliff of Coroglio (Posillipo hill, Naples, Italy; Figure 1) is located within the active volcanic caldera of Phlegrean Fields, and has been selected in order to include several factors, as well as slope morphology, exposure of coastal cliffs to marine erosion (due to the wave action, the wind abrasion, the thermal excursions, etc.), type of geological substratum and anthropogenic pressure [9, 10].

The experimental results have shown that the sensing optical fiber cable is able to detect the formation, the evolution and the location of fractures on tuff rocks.

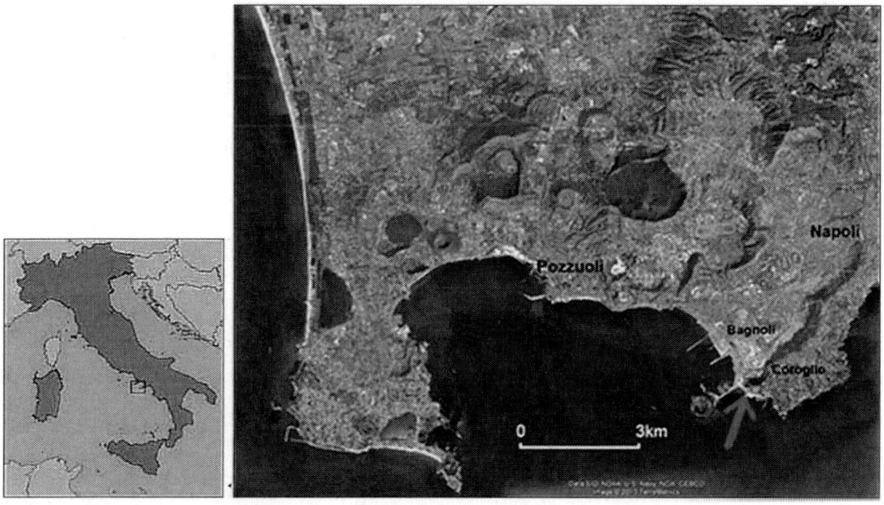

Figure 1. Location map showing the Phlegrean Fields area with the Coroglio site location (red arrow).

Figure 2. Frontal view of the Coroglio cliff. The insert shows the test site position at the base of the slope. The yellow line shows the site of the experiment.

Geological Characteristics of the Test Site

The coastal cliff of Coroglio is located in the Phlegrean Fields active volcanic area (Figure 1) in the western sector of Naples town. It has been selected as representative of a tuff coastal cliff in a densely populated area, associated with significant natural hazards [10]. The Coroglio cliff is 140 m high and 250 m wide (Figure 2).

The uppermost part of the outcropping succession is characterized by an about 30 m thick loose pyroclastic deposits (Figure 3). The coastal cliff is characterized by the outcrop of the upper member of the Neapolitan Yellow Tuff, a lithified ignimbritic deposit dated at ca. 15 ka BP [11]. The exposed rocks are formed by alternating coarse-grained matrix-supported breccia, thin-laminated lapilli beds and massive ash layers. The deposit overlies the older tuff cone of Trentaremi (dated ca. 22.3 ka BP [12, 13]), consisting of slightly welded to welded, whitish-yellow, pumiceous coarse-grained fragments in a sandy ash matrix and of lapilli beds [14]. At the base of the cliff slope talus breccia and beach deposit also occur (Figure 3). The upper part of the cliff has been reinforced by steel bars anchored and bolted to the rock and by wire mesh and steel cable network applied to the tuff wall, realized in the last 15 years after relevant rock falls occurred between the years 1985 and 1990.

The volcaniclastic succession cropping out at the Coroglio cliff is characterized by a complex system of structural discontinuities and fractures [9], mostly steep and planar with highly variable density, with well-developed NE–SW and NW–SE directions and subordinate N-S and E-W trends.

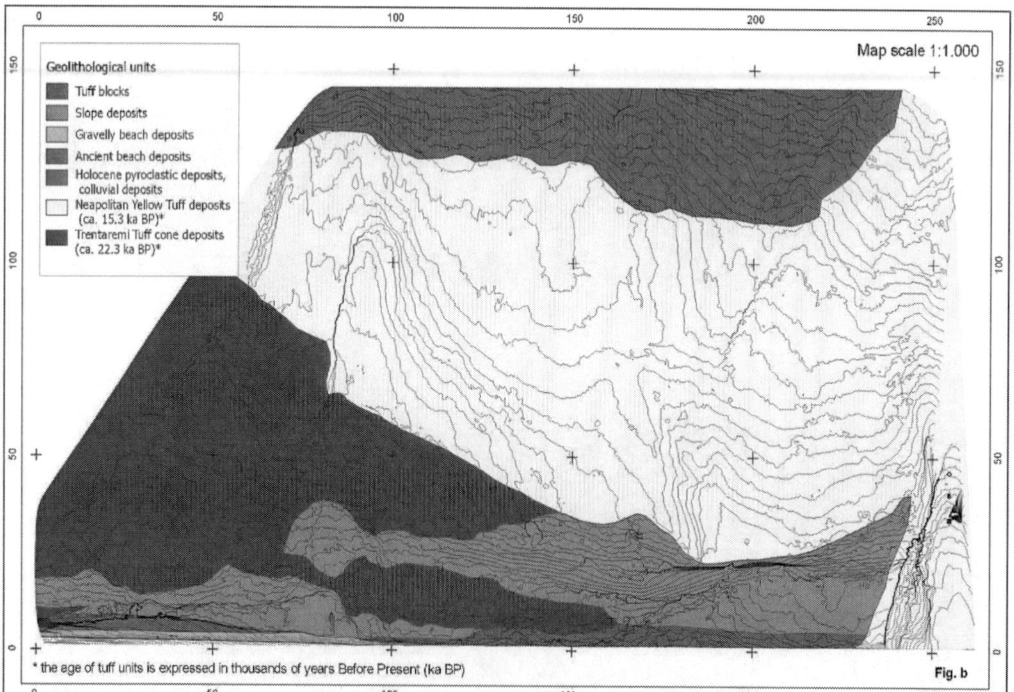

Figure 3. Geological map of Coroglio cliff [9]. The age of tuff units is expressed in thousands of years Before Present (ka BP). Contour lines interval is 0.25 m.

METHODS

Interrogation System

Stimulated Brillouin scattering in optical fibers allows distributed measurements of strain and temperature over large distances and with high spatial resolution. SBS sensors rely on the dependence of the so-called Brillouin frequency shift on the strain and temperature conditions of the fiber. In BOTDA sensors, the Brillouin frequency shift profile along the fiber is retrieved by recording the interaction between a pulsed pump beam and a counter-propagating continuous wave probe as a function of time, while scanning the pump-probe spectral shift over a typical ~300MHz frequency range. Any deviation of the local Brillouin frequency shift value from a reference measurement is a signature of strain or temperature change (Figure 4).

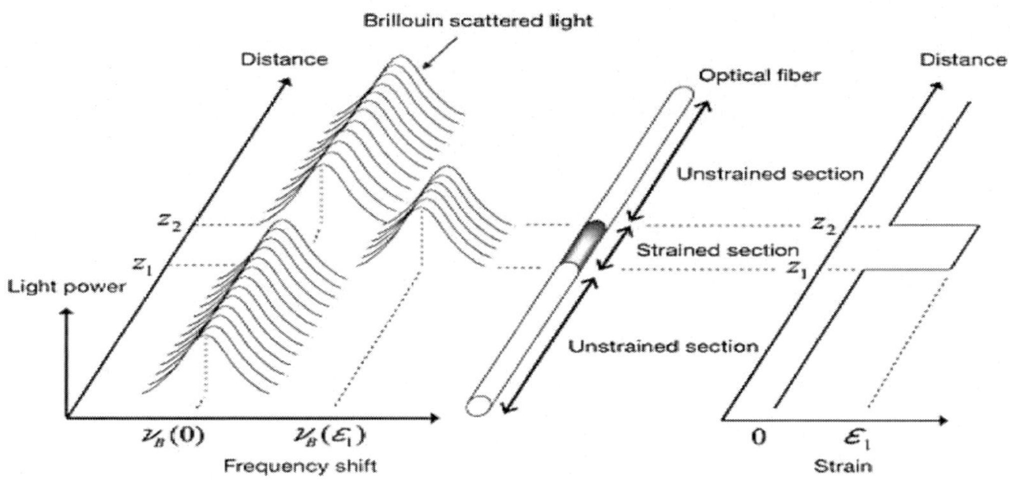

Figure 4. Principle diagram of BOTDA technology [18].

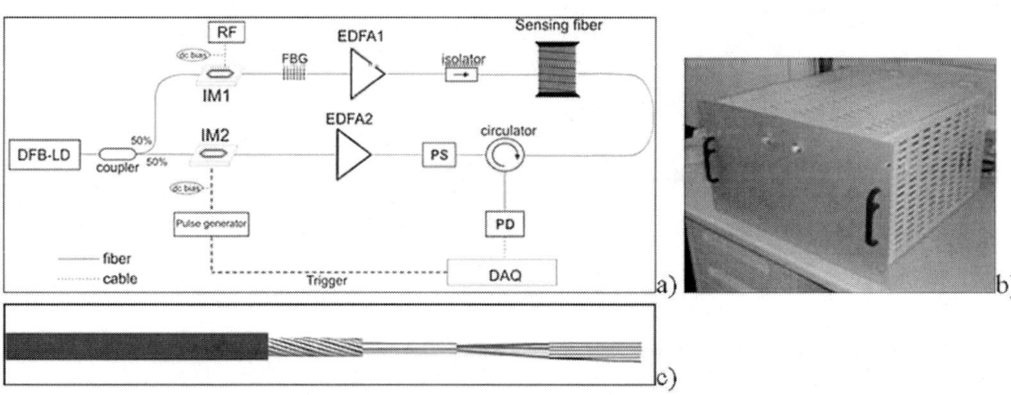

Figure 5. a) Scheme of the optoelectronic measurement setup; b) Picture of the optoelectronic measurement unit; c) used sensing cable.

The BOTDA setup used for the strain measurements is shown in Figure 5a. The light from a distributed feedback laser diode (DFB-LD) operating at 1.55 μm wavelength is split into two distinct arms to generate the pump and probe beams. The lower branch in the figure shows the generation of the pulsed pump, while the upper branch shows the generation of the probe signal through double-sideband suppressed-carrier modulation and a fiber Bragg grating (FBG) that selects the sideband at lower frequency (Stokes component). Finally, a polarization scrambler (PS) is employed to average out the Brillouin gain fluctuations associated to variations of state of polarization of the interacting beams [15].

The measurement unit (Figure 5b) allows to measure strain and temperature profiles up to tens of kilometers with a strain accuracy < 20 με, a temperature accuracy < 1°C and a spatial resolution of 1 m. As the sensing cable, a standard low-loss single mode optical fiber with tight-buffer of 900μm has been chosen (Figure 5c). It should be emphasized that, due to the distributed nature of the sensor, no preliminary information about the possible locations of displacement in the rocks is required in advance.

Figure 6. Sensing fiber cable path with fixed segment of the cable (Str-1 to Str-9 lines in yellow) and fracture systems in the tuff (S1-F1-F2-F3-F4 dashed orange and red lines). The chisel insertion point is In-1 (orange circle).

Sensor Deployment

The optical fiber cable has been deployed in September 2013 at the base of the tuff slope by spot gluing the cable with epoxy adhesive (Figure 6). In order to assess the validity of the proposed approach, a few existing cracks have been artificially enlarged by progressively driving chisels of increasing size into a selected fracture and the sensor has been used to identify the cracks and to follow their growth.

EXPERIMENTAL RESULTS AND DATA ANALYSIS

The tuff cracks have been detected with a localization of 1-m spatial resolution and a signal processing technique based on spectrum shape analysis and multiple peaks detection [16]. Fractures smaller than the sensor spatial resolution generate additional peaks in the measured spectrum. The spectrum is characterized by unstrained and strained peak frequencies. The strain information is extracted using a multiple distribution fitting algorithm, in which the initial conditions, such as the unstressed fiber Brillouin shift and the Brillouin spectrum line width at each position, are carefully set. The Brillouin gain spectrum (BGS) measured along the sensing fiber at position z = 14.3 m (Str-3) is shown in Figure 7.

The strain variation was recorded with seven measurements during the forcing of the crack (F1) with a chisel in the insertion point In-1, located between Str-3 and Str-4 (Figure 6).

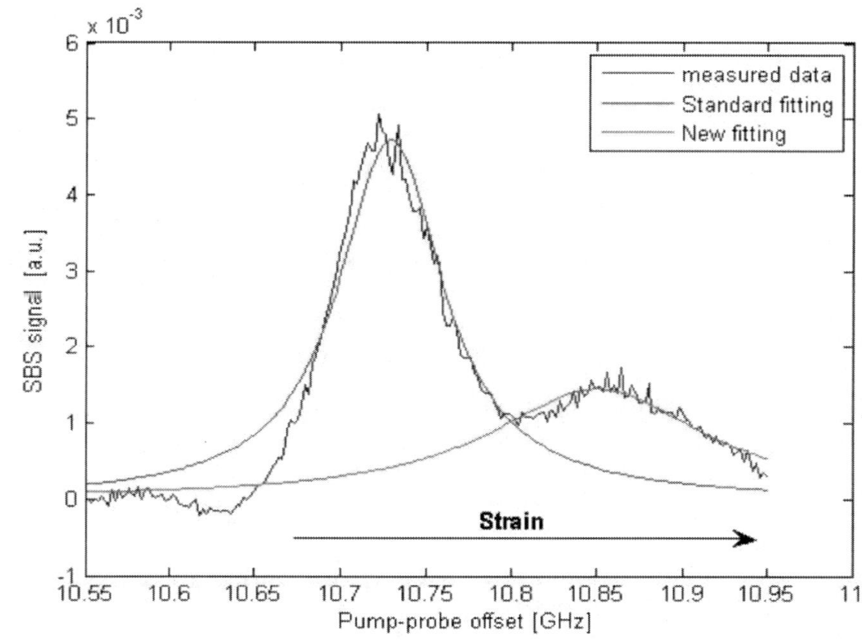

Figure 7. Brillouin gain spectrum measured at the position z=14.3 m (Str-3) of the sensing fiber.

Seven different measurement positions have been localized along the optical fiber (Str-1, 2, 3, 5, 7, 8, 9 in Figure 5). A reference segment of 0.75 meters across each measurement position has been considered in order to measure the maximum strain values. The segment length was calibrated to compensate the possible coupling problems between the fiber and the rocks. The uniformed maximum strain (UMS) has been used to evaluate the strain evolution for each measurement segment (Seg. 1, 2, 3, 5, 7, 8, 9 in Figure 8) during the seven measurements. Each curve describes the variation of the maximum value of strain in each measurement position for all the measures (Figure 8).

It can be observed that in the segment 3 (Seg. 3 in Figure 8), corriponding to the measurement position 3 (Str.-3 in Figure 6), the widening of the crack F1 is promptly recorded by the strain variation when chivel is progressively inserted in the infission point In-1 (Figure 6), because the position 3 is the nearest. The more distant segments (seg. 1, 2, 5, 7 in Figure 8), corresponding to the measurement positions 1, 2, 5, 7 (Str.- 1, 2, 5, 7 in Figure 6), record the crack widening with a delayed and lowered intensity.

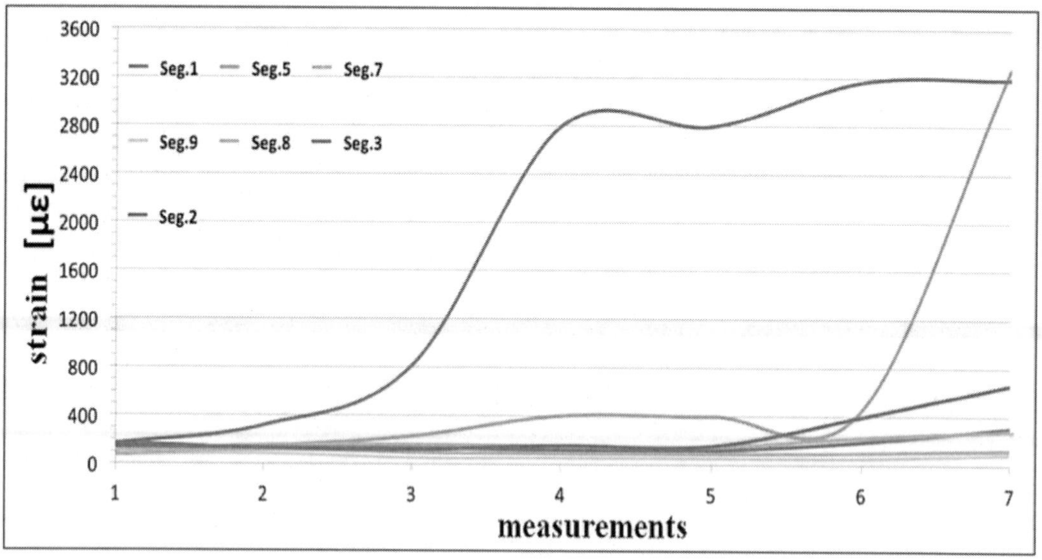

Figure 8. Strain variations for each measurement position (lines) for all the seven measurements.

Conclusion

An optical fiber strain-sensing technique, based on Brillouin optical time domain analysis (BOTDA), was used in a field test application to monitor the crack opening in a rock slope made by volcanic tuff. The obtained results clearly show how the sensing optical fiber system is able to detect the formation of cracks, identify their location along the slope and follow their evolution.

This field monitoring test has allowed to verify the proper efficiency of the designed optical fiber detection system in a controlled natural context. It has been preparatory for the planned application to the effective geological monitoring of the Coroglio tuff cliff [17]. As a matter of fact, a new larger scale installation has been performed since June 2015, aimed to a permanent monitoring of the tuff cliff subject to rock failures hazard.

ACKNOWLEDGMENTS

Financial support for this research was provided by the Programma Operativo Nazionale (PON) –Ricerca e Competitivita' 2007–2013 funded by the Italian Ministry of University and Research – Project MONICA [grant PON01_01525].

REFERENCES

[1] López-Higuera, J. M., Cobo, L. R., Incera, A. Q., Cobo, A., (2011). Fiber Optic Sensors in Structural Health Monitoring. *Journal of Lightwave Technology* 29, 586-608.

[2] Minardo, A., Bernini, R., Zeni, L., (2011). Numerical analysis of single pulse and differential pulse-width pair BOTDA systems in the high spatial resolution regime. *Optics Express* 19, 19233-19244.

[3] Bernini, R., Minardo, A., Ciaramella, S., Minutolo, V., Zeni, L., (2011). Distributed strain measurement along a concrete beam via stimulated Brillouin scattering in optical fibers. *International Journal of Geophysics*, vol. 2011, art. ID 710941, 1-5. doi: 10.1155/2011/710941.

[4] Minardo, A., Bernini, R., Amato, L., Zeni, L. (2012). Bridge monitoring using Brillouin fiber-optic sensors. *IEEE Sensor Journal* 12 (1), 145-150.

[5] Terzaghi, K., (1950). Mechanism of landslides. In: S. Paige Ed., *Application of geology to engineering practice* (pp. 83-123). New York: Geological Society of America.

[6] Varnes, D.J., (1983). Time-deformation relations in creep to failure of earth materials. *Proceedings of the 7th Southeast Asian Geotechnical Conference*, Hong Kong, 107-130.

[7] Fukuzono, T., (1990). Recent studies on time prediction of slope failure. *Landslide News* 4, 9-12.

[8] Voight, B. (1989). A relation to describe rate-dependent material failure. *Science* 243, 200-203.

[9] Matano, F., Iuliano, S., Somma, R., Marino, E., del Vecchio, U., Esposito, G., Molisso, F., Scepi, G., Grimaldi, G. M., Pignalosa, A., Caputo, T., Troise, C., De Natale, G., Sacchi, M., (2015). *Journal of Maps.* http://dx.doi.org/10.1080/17445647.2015.1028237.

[10] Somma, R., Matano, F., Marino, E., Caputo, T., Esposito, G., Caccavale, M., Carlino, S., Iuliano, S., Mazzola, S., Molisso, F., Sacchi, M., Troise, C., De Natale, G. (2014). Application Of Laser Scanning For Monitoring Coastal Cliff Instability In The Pozzuoli Bay, Coroglio Site, Posillipo Hill, Naples. In G. Lollino, A. Manconi, F. Guzzetti, M. Culshaw, P. Bobrowsky, & F. Luino (Eds.), *Engineering geology for society and territory* (Vol. 5, Chap. 133, pp. 687–690). Cham: Springer International.

[11] Deino, A.L., Orsi, G., de Vita, S., Piochi, M., (2004). The age of the Neapolitan Yellow Tuff caldera-forming eruption (Campi Flegrei caldera-Italy) assessed by 40Ar/39Ar dating method. *J. Volcanol. Geotherm. Res.* 133, 157– 170.

[12] Cole, P. D, Perrotta, A., Scarpati, C. (1994). The volcanic history of the southwestern part of the city of Naples. *Geological Magazine,* 131, 785-799.

[13] Perrotta, A., Scarpati, C., (2012). I vulcani di Napoli: guida geologica all'escursione alla villa di Pollione di Posillipo. In: Aracne Ed., Scienze Naturali e Archeologia (pp. 1-7), Rome.

[14] Lirer, L., (2011). I Campi Flegrei. Storia di un campo vulcanico. *Quaderni dell'Accademia Pontaniana* 57, 1-175.

[15] Minardo, A., Coscetta, A., Pirozzi, S., Bernini, R., Zeni, L. (2012). Modal analysis of a cantilever beam by use of Brillouin based distributed dynamic strain measurements. *Smart Mater. Struct.* 21, 125022.

[16] Ravet, F., Briffod, F., Glišić, B., Niklès, M., Inaudi, D. (2009). Submillimeter Crack Detection With Brillouin-Based Fiber-Optic Sensors. *IEEE Sensors Journal* 9(11), 1391-1396.

[17] Sacchi, M., Caputo, T., Esposito, G., Somma, R., Matano, F., Caccavale, M., del Vecchio, U., Fortelli, A., Grimaldi, G. M., Iuliano, S., Latte Bovio, G., Marino, E., Di Meglio, C., Minardo, A., Pignalosa, A., Terracciano, R., Troise, C., Zeni, L., De Natale, G. (2015). Sistema integrato di monitoraggio (TCM - Tuff Cliff Monitoring) dei fenomeni di dissesto lungo il costone tufaceo di Coroglio (Napoli). INGV, Rapporti Tecnici, 309, 72 pp., ISSN 2039-7941.

[18] Breteler, R., van der Tol, J. M., Felicetti, M., Sasbrink, B., Smit, M. K. (2011). Photonic integrated brillouin optical time domain reflection readout unit. *Opt. Eng.*, 50(7), 071111-071111-8.

About the Editor

Gemma Aiello, PhD
Researcher
Institute of Marine and Coastal Environment,
National Research Council of Italy,
Calata Porta di Massa, Porto di Napoli, Napoli, Italy
Email: gemma.aiello@iamc.cnr.it

Dr. Gemma Aiello was born in Aversa (CE), Italy on 24 October 1964. In 1989 she graduated in Geological Sciences at University of Naples "Federico II". In 1993 she earned a PhD in Sedimentary Geology at the University of Naples "Federico II", Department of Earth Sciences, Faculty of Geological Sciences. She completed a two year post-doctoral fellowship at the University of Naples "Federico II", a CNR-CEE fellowship and several contracts at the Research Institute "Geomare Sud", CNR, Naples, Italy. From 1998 to 2001 she was a contract researcher at the aforementioned Institute. Since 2001 she has been a full time researcher at the National Research Council of Italy, Institute of Marine and Coastal Environment (IAMC), Calata Porta di Massa, Porto di Napoli, Napoli, Italy. Dr. Aiello has twenty-years experience in the field of marine geology and geophysics, participating in numerous research projects of CNR (CARG, Vector, Centri Regionali di Competenza, etc.). She has carried out didactic activity as tutor of degree and PhD theses and teacher of geology at the University Parthenope of Naples and CNR. She has participated in more than twenty oceanographic cruises, acquiring experience in the acquisition of reflection seismics and Multibeam bathymetry. Her scientific activity is documented by many national and international publications and by abstracts in national and international conferences in the field of Earth Sciences. She has participated in the CARG research project as geological surveyor and Director of geological survey of marine geological maps. Dr. Aiello has participated in the TESSA project (Operational Oceanography) and coordinated the didactic activities of the OTTIMA research project at the CNR as scientific responsible during the 2013, 2014 and 2015, forming 50 fellowship researcher and technologists in the field of marine sciences.

INDEX

A

air quality, 149
air temperature, 160
amplitude, 3, 5, 12, 17, 20, 62, 72, 89, 105, 125, 127, 128, 129, 130, 132, 133, 144, 145, 172
anisotropy, 17
aquifer, viii, 1, 10, 23, 25, 27, 39, 40, 42, 43, 49, 56
atmosphere, 10, 124, 129, 141, 145

B

background noise, 48, 115
banks, 71, 72, 78, 81, 94, 100, 140
Bayesian estimation, 121
bedding, 107
benchmarks, 87
bioremediation, 42
boreholes, 4, 13, 27, 29, 50
Brillouin Optical Time Domain Analysis (BOTDA), vii, ix, 179, 180, 181, 183, 184, 186, 187

C

cables, 7, 9, 11, 12, 49, 50, 56
calibration, 9, 11, 146
carbon dioxide (CO2), 36, 117
case study(ies), viii, 1, 2, 35, 36, 38, 44, 58, 115, 177
Central Europe, 36
changing environment, viii, 47
chemical characteristics, 104
chemical reactions, 7
chlorophyll, 177
circadian rhythm, 124, 125
circulation, vii, ix, 123, 125, 133, 137, 138, 139, 140, 142, 144, 146, 156, 166, 172, 174, 175, 176
climate, 149, 178
clusters(ing), 88, 89
coal, 39
coastal management, 50
conductivity, 3, 11, 12, 13, 20, 23, 27, 28, 29, 146
conservation, 2, 143
construction, viii, 1, 2, 8, 27, 28, 31, 47, 48, 50, 53, 80, 149, 180
contaminant, 23, 49
contamination, 10, 23, 37
continental shelf, vii, 72, 94, 95, 96, 97, 98, 101, 105, 109, 138, 141
Coroglio, v, vii, x, 179, 181, 182, 186, 187, 188
Coroglio coastal cliff, vii
correction factors, 19
cracks, vii, 88, 181, 185, 186
Croatia, 111
crop, 27, 105
crust, viii, 1, 118, 120, 121
crystalline, 37

D

data collection, 15, 17, 20, 23, 41
data processing, viii, 1, 2, 3, 5, 13, 14, 21, 56, 86, 110
data set, 23, 24, 148
decay, 7, 12
decomposition, 149, 172, 173, 176, 177
deconvolution, 63
deformation, 87, 95, 102, 103, 112, 116, 118, 180, 187
degradation, vii, 1, 23
deposition, 95, 107, 109
deposits, 7, 23, 27, 45, 53, 66, 69, 70, 71, 74, 75, 77, 78, 79, 82, 86, 88, 91, 95, 102, 105, 106, 107, 110, 118, 182
depression, 69, 75, 77, 79, 95, 102, 105

depth, 3, 4, 5, 6, 7, 10, 11, 12, 13, 14, 15, 18, 20, 21, 26, 27, 31, 34, 41, 48, 50, 54, 62, 65, 68, 69, 72, 74, 80, 81, 86, 88, 90, 98, 107, 109, 140, 144, 152, 154, 157
detection, viii, x, 1, 2, 7, 9, 10, 11, 28, 32, 36, 37, 39, 40, 41, 58, 80, 89, 124, 179, 180, 185, 186
detection system, 186
dielectric constant, 12, 55
diffusion, 7, 144, 150, 157, 173
diffusion time, 157
diffusivity, 145
dipoles, 12
discharges, 81
disequilibrium, 56
dislocation, 173
dispersion, 4, 16, 49, 54
displacement, 80, 87, 180, 184
distribution, viii, 6, 14, 17, 20, 21, 34, 42, 47, 49, 56, 77, 86, 87, 88, 89, 90, 91, 93, 102, 103, 104, 119, 120, 127, 130, 144, 147, 149, 150, 151, 185
drainage, 10, 21, 87
drought, 42
dykes, 6, 68, 72, 95, 101

E

early warning, vii, x, 179
earthquakes, 71, 87, 88, 89, 114
ecosystem, 48, 133
electric current, 6, 7, 11, 55
electric field, 55, 89
electrical conductivity, 12
electrical properties, 6, 12, 13, 89
electrical resistance, 38, 58
electrodes, 6, 7, 15, 17, 23, 55, 56, 62, 63, 64, 81, 83
electromagnetic, vii, 2, 3, 11, 12, 13, 25, 27, 28, 29, 31, 40, 43, 58, 62, 88, 180
electromagnetic waves, 11, 12, 13
emission, 79, 88
Empirical Orthogonal Functions (EOF), vii, ix, 137, 138, 139, 147, 148, 149, 166, 168, 170, 171, 172, 173, 177
energy, 5, 7, 12, 14, 31, 62, 63, 64, 75, 76, 78, 80, 81, 84, 85, 103, 124, 128, 131, 142, 145, 160, 161, 176
engineering, 1, 3, 4, 8, 9, 12, 21, 27, 29, 34, 35, 36, 39, 41, 44, 48, 49, 50, 176, 187
environmental aspects, 48
environmental change, 23
environmental degradation, vii, 1
environmental impact, 117
environmental management, 41
environments, 44, 48, 58
equipment, 3, 7, 14, 15
erosion, viii, 47, 48, 181
Euclidean space, 149
European Commission, 48, 58
evaporation, 23, 42, 142, 146, 171, 172
evolution, vii, 50, 71, 74, 79, 103, 104, 106, 110, 112, 113, 114, 115, 119, 138, 181, 186
excavations, 29
explosives, ix, 61, 62, 64, 80

F

facies, 91, 94, 95, 104, 105, 107, 110, 112
fault zones, viii, 1, 27
Federal Highway Administration (FHWA), 36, 43
fiber optic sensors, 179
finite element method, 143
flora and fauna, 62
flow field, 125
forecasting, 175, 177
formation, ix, 7, 11, 13, 31, 74, 75, 77, 78, 79, 84, 102, 103, 110, 124, 171, 175, 179, 181, 186
fractures, ix, 35, 88, 102, 179, 180, 181, 182
freshwater, viii, 40, 48, 49, 55, 142, 145
friction, 125, 126

G

geology, viii, 13, 17, 27, 35, 40, 41, 47, 48, 49, 70, 71, 77, 98, 180, 187
geometry, viii, 8, 13, 14, 30, 47, 49, 50, 55, 56, 63, 90, 91, 104, 105, 141
geophysics, vii, viii, 1, 2, 8, 11, 12, 13, 35, 36, 37, 40, 41, 42, 61, 106, 111, 117
GIS, 22, 38, 87
GPS, 9, 86, 96, 180
gravitational field, 38
gravity, 2, 3, 8, 9, 10, 15, 18, 31, 36, 37, 38, 39, 40, 41, 42, 87, 88, 89, 104, 112, 114, 118, 125, 138
Greece, vii, 1, 21, 22, 27, 31, 32, 36, 37, 38, 39, 40, 41, 42, 43, 44, 45
groundwater, 2, 6, 7, 12, 23, 25, 27, 35, 39, 40, 42, 43, 49, 56, 89, 180

H

Hawaii, 118
hazards, 2, 21
heat loss, 171
heavy metals, 2, 11, 22
hemisphere, 142

historical data, 70, 87
Holocene, 22, 69, 91, 94, 117, 120
hydrothermal activity, 74, 98

I

IDA, 110
identification, viii, 7, 32, 48, 71, 85, 89, 90, 138
imagery, 38, 175
image(s), 3, 4, 5, 7, 15, 23, 24, 27, 29, 32, 33, 34, 37, 42, 48, 52, 54, 58, 68, 71, 88, 89, 115, 141
industrial emissions, 10, 40
industry, 3, 4, 9, 34
inertia, 141
inferences, 111, 119
infrastructure, 35
islands, vii, viii, 51, 61, 62, 71, 74, 77, 102, 109, 120
isotope, 114, 119
Italy, v, vii, x, 45, 47, 58, 61, 62, 65, 81, 82, 86, 111, 112, 113, 114, 115, 116, 117, 118, 119, 120, 121, 137, 147, 179, 181, 187

J

Japan, vii, 123, 133, 135

K

Kuroshio, vii, ix, 123, 127, 133, 134

L

land utilization, vii, 1
landfills, 23
landscape, 32
landslide monitoring, 179, 180
landslides, viii, 1, 2, 89, 110, 114, 180, 187
Late Pleistocene, 91, 92, 105, 116
limestone, 27, 29, 82
linear systems, ix, 123, 124

M

magnetic field, 10, 11, 18, 19, 41, 65, 66, 67, 71, 72, 86, 90, 92, 94
magnetic materials, 62
magnetic properties, 10, 21, 86, 96
magnetism, 10, 120
magnetization, 10, 21, 92, 96, 114
magnetosphere, 65
magnitude, 5, 7, 11, 65, 88
main belt, 90
Malaysia, 44
management, viii, 27, 35, 47, 48, 50, 56
mapping, vii, viii, 1, 4, 6, 12, 19, 27, 28, 31, 35, 36, 40, 42, 43, 50, 53, 64, 84, 94, 106
marine environment, 14, 50, 53, 105
MEAS (Multispot Extended Array Sparker), vii, ix, 61, 62, 64, 81, 82
measurements, 3, 5, 6, 7, 10, 11, 14, 17, 18, 19, 20, 23, 27, 28, 29, 32, 38, 39, 43, 55, 56, 64, 65, 87, 89, 90, 94, 96, 98, 112, 147, 150, 180, 183, 184, 185, 186, 188
Mediterranean, vii, ix, 78, 113, 137, 138, 139, 140, 143, 144, 173, 174, 175, 176, 177, 178
melting, 142, 180
metals, 11, 21, 98
meter, 12, 13, 81
methodology, 28, 41, 50, 106, 143
Mexico, 43
microgravity, 3, 8, 9, 39, 40
migration, 15, 49, 88, 113
Miocene, 82
mixing, 49, 106, 125, 126, 144, 154, 173
modelling, 116, 141, 177
models, viii, 6, 14, 17, 27, 28, 29, 30, 31, 33, 34, 43, 47, 50, 53, 55, 56, 68, 69, 81, 87, 88, 90, 102, 112, 121, 125, 138, 143, 144, 174, 176, 178
momentum, 124, 144, 145
morphology, 7, 10, 181

N

Naples, v, vii, viii, x, 47, 61, 62, 64, 65, 66, 70, 71, 74, 77, 78, 79, 80, 81, 82, 83, 84, 85, 86, 87, 90, 91, 94, 95, 98, 101, 102, 106, 107, 109, 110, 111, 112, 114, 116, 117, 118, 119, 120, 179, 181, 182, 187
native species, 56
natural hazards, 182
natural resources, vii, 1, 2, 48
Nicaragua, 43
NOAA, 146, 177
non-linear system, ix, 123, 124
North Africa, 138

O

oceanic circulation, ix, 123, 125, 133, 142
oceanic currents, 125, 133
oceanic oscillation, v, vii, 133
oceans, 176

olive oil, 23, 37, 42
optical fiber, vii, ix, 179, 180, 181, 183, 184, 185, 186, 187
oscillation, vii, ix, 123, 124, 125, 130, 133, 167
oscillators, 124

P

Pacific, ix, 123, 124, 127
physics, 149, 150, 173
Pliocene, 74
polarity, 7, 17, 94
polarization, 2, 7, 37, 184
pollution, viii, 6, 10, 21, 34, 36, 47, 48
polyphenols, 23
ponds, 23, 42
porosity, 3, 55
porous media, 58
Portugal, 38
power plants, 22
precipitation, 142, 146, 177, 180
pressure gradient, 142, 144, 176, 177
Principal Components Analysis (PCA), 147
Procida, vii, viii, 50, 51, 53, 58, 59, 61, 62, 69, 71, 73, 74, 77, 78, 82, 102, 105, 108, 109, 113, 115, 119, 120
propagation, 5, 12, 13, 90, 176
public health, 2, 35
pure water, 145

Q

quasi-geostrophic, ix, 123, 125

R

radar, 2, 13, 27, 31, 41
radiation, 63, 142, 145, 177
rainfall, 87
red noise, 125, 127, 129, 133
reflection techniques, vii
resources, 25, 35, 56, 82, 83
rock slope failures, 179
Romania, 58
rotation axis, 103
runoff, 146

S

salinity, 55, 56, 140, 142, 144, 145, 146, 151, 152, 154, 156, 157, 164, 166, 170, 171, 172, 173
saltwater, viii, 48, 49, 55, 56
sea level, 72, 79, 94, 118, 120, 124, 177
seasonal variation, 56, 127, 128, 129, 130, 133, 171
sedimentation, 72, 91, 94, 116
sediments, viii, 10, 21, 27, 41, 47, 48, 50, 62, 71, 81, 82, 83, 86, 92, 101, 107
seismic data, vii, viii, ix, 5, 14, 15, 29, 38, 47, 48, 49, 50, 61, 62, 65, 69, 81, 86, 89, 116
seismic refraction, vii, 4, 27, 32, 44, 49, 50
sensors, vii, ix, 11, 13, 49, 51, 58, 65, 86, 146, 179, 180, 183, 187, 184, 185
shoreline, 49, 50, 98, 143
Sicily Channel, v, vii, ix, 137, 138, 140, 141, 143, 146, 160, 167, 172, 175, 176, 177
signals, ix, 2, 4, 12, 20, 64, 85, 89, 115, 116, 123, 124, 128, 132, 133, 146, 166, 171, 172
simulations, 138, 144, 146, 160, 161, 174, 175, 176
sinkholes, viii, 1, 2, 4, 6, 39
Spain, 38
stochastic resonance, vii, ix, 123, 124, 125, 129, 131, 132, 133, 134, 135
storms, 19, 138
stratification, 53, 154, 173
stress, 89, 102, 103, 124, 125, 126, 127, 129, 145, 160, 166, 177
structural characteristics, 116
structure, viii, 1, 4, 7, 11, 12, 14, 29, 48, 64, 67, 68, 73, 77, 82, 86, 88, 90, 91, 93, 94, 96, 111, 113, 115, 120, 121, 124, 125, 126, 147, 159, 173, 174, 175, 180
surface layer, 13, 23, 160, 175
surface properties, ix, 137
surface structure, 2, 32, 36, 113
survey design, 14, 49, 50
susceptibility, 3, 10, 11, 18, 21, 22, 40, 87, 96, 181
synchronization, v, vii, ix, 123, 124, 125, 130, 131, 133, 134, 135
synchronize, 124

T

techniques, vii, viii, 1, 2, 3, 4, 6, 8, 10, 13, 19, 21, 22, 29, 31, 32, 34, 35, 37, 38, 39, 40, 44, 47, 48, 61, 63, 66, 96, 110, 115, 117, 143
technological advances, 65, 110
technologies, vii, viii, 61, 62, 64, 66, 79, 80, 83, 106, 110, 111
technology, 63, 83, 183
temperature, 10, 11, 21, 55, 117, 142, 143, 144, 145, 146, 150, 151, 152, 154, 156, 157, 159, 160, 167, 168, 171, 172, 176, 177, 180, 183, 184
temporal variation, 56, 89
terraces, 109

thermal energy, 103
tides, 18
time periods, 124
time series, 132, 138, 149, 150, 156, 157, 161
total energy, 127, 128, 129, 130, 131, 132
toxic substances, 35
toxic waste, 9
trace elements, 104
transducer, ix, 61, 65, 81
transformations, 21, 36
transport, viii, 37, 47, 48, 58, 124, 138, 140, 144, 155, 160, 172, 175, 176
triggers, 62, 103
tuff, ix, 50, 53, 72, 74, 77, 79, 106, 179, 181, 182, 184, 185, 186
tuff cliff, x, 179, 186
turbulence, 125, 144, 176

U

U.S. Department of Commerce, 112
U.S. Geological Survey (USGS), 36, 58, 120
UNESCO, 144
urban areas, vii, 1, 13, 36, 37, 42

V

vector, 65, 148
vegetation, 10, 11, 13
velocity, 3, 4, 5, 6, 9, 12, 13, 14, 16, 30, 31, 34, 48, 49, 53, 54, 68, 72, 74, 88, 89, 90, 107, 113, 115, 117, 120, 121, 127, 141, 142, 143, 144, 145, 164, 166
Ventotene islands, vii
viscosity, 126, 127, 142, 144

W

wastewater, 37, 42
water, viii, 4, 7, 12, 18, 27, 39, 40, 42, 48, 49, 50, 53, 55, 56, 57, 58, 62, 64, 80, 81, 84, 85, 89, 94, 98, 117, 124, 138, 140, 142, 145, 146, 154, 157, 171, 173, 174, 175, 176
water chemistry, 89
water resources, viii, 48
watershed, 25
wave propagation, 178
wavelengths, 4
wavelet analysis, 177
wells, 71, 75, 77, 89
western boundary current, vii, ix, 123, 127, 133
wind turbines, 4, 29, 31